# THE TRUCK BOOK

# THE
# TRUCK
# BOOK

## THE DEFINITIVE
## VISUAL HISTORY

Penguin
Random
House

**Produced for DK by**
**Dynamo Limited**
1 Cathedral Court, Southernhay East,
Exeter, EX1 1AF

**DK UK**
**Senior Editor** Chauney Dunford
**Senior Art Editor** Gillian Andrews
**Photographer** Gary Ombler
**Managing Editor** Gareth Jones
**Senior Managing Art Editor** Lee Griffiths
**Production Editor** Jacqueline Street-Elkayam
**Production Controller** Laura Brand
**Jackets Design Development Manager** Sophia MTT
**Art Director** Karen Self
**Associate Publishing Director** Liz Wheeler
**Publishing Director** Jonathan Metcalf

**DK INDIA**
**Senior Art Editor** Ira Sharma
**Project Editors** Nandini D. Tripathy, Hina Jain
**Art Editor** Aanchal Singal
**Assistant Art Editors** Sulagna Das, Ananya Gyandhar
**Managing Editor** Soma B. Chowdhury
**Senior Managing Art Editor** Arunesh Talapatra
**Jacket Designer** Juhi Sheth
**Senior Jackets Coordinator** Priyanka Sharma Saddi
**DTP Designers** Deepak Mittal, Mrinmoy Mazumdar,
Vikram Singh, Mohd Rizwan
**DTP Coordinator** Vishal Bhatia
**Hi-res Coordinator** Neeraj Bhatia
**Production Manager** Pankaj Sharma
**Pre-production Manager** Balwant Singh
**Creative Head** Malavika Talukder

**Contributors** Dan Parton, Ellen Voie, Gareth Jones,
Giles Chapman, Wes Nicholson

**Consultant** Ashley Hollebone

First published in Great Britain in 2024 by
Dorling Kindersley Limited
DK, One Embassy Gardens, 8 Viaduct Gardens,
London, SW11 7BW

The authorised representative in the EEA is
Dorling Kindersley Verlag GmbH. Arnulfstr. 124,
80636 Munich, Germany

Copyright © 2024 Dorling Kindersley Limited
A Penguin Random House Company
10 9 8 7 6 5 4 3 2 1
001–336906–Apr/2024

A CIP catalogue record for this book
is available from the British Library.
ISBN: 978-0-2416-3480-6

Printed in China

www.dk.com

# Contents

## 1890–1919: THE EARLY YEARS

At the beginning of the age of motoring, pioneering
engineers commonly built trucks on existing car chassis.
Powered by steam, petrol, and even electricity, early
trucks operated alongside horse-drawn carriages. The
first commercial vehicles became available around the
turn of the century, but the demand for military transport
during World War I accelerated truck production, and
jumpstarted a wider market for trucks thereafter.

## 1920–1938: STYLE AND SUBSTANCE

The Roaring Twenties saw a diverse range of commercial trucks emerge that echoed the bold, modern style of the day. As assembly-line mass production became standard, diesel also became the preferred fuel for many trucks.

## 1939–1959: IN WAR AND PEACE

During World War II, most truck production was dedicated to the war effort, including flat-pack vehicles shipped for assembly at front lines around the world. Many military models had civilian afterlives, contributing to rebuilding efforts, and laying the groundwork for a powerful new generation of trucks for every purpose.

## 1960s: THE GOLDEN AGE

This decade saw a wave of developments in truck design that combined a greater focus on styling, driver comfort and safety, and fuel efficiency. As road networks expanded, trucks could travel further and faster.

## 1970s: LIVING THE DREAM

By the 1970s, trucking had not only developed a strong community, but also become a pervasive theme in popular culture at large. Trucks and truckers became the stars of many films and TV shows at this time, inspiring viewers with life on the road.

## 1980–1999: TURBO POWER

As pumped-up trucks began gracing walls as pin-up posters, monster trucks became arena entertainment. Manufacturers launched muscle trucks and performance models, making trucks bigger, tougher, and faster.

## 2000–TODAY: KEEP ON TRUCKIN'

Trucks have become ever more refined in the 21st century, engineered to maximize power, capacity, and fuel efficiency, and fitted with advanced systems to make vehicles safer. As the world looks towards a zero-emissions future, manufacturers are investing heavily in battery-electric and hydrogen-cell technology, while leading companies develop prototypes of driverless trucks.

## HOW TRUCKS WORK: TRUCK TECHNOLOGY

This section covers the basics of truck engineering, from the engines that power them, to the chassis that support them. It provides a brief overview of suspension systems and a look at current cab design.

# What is a Truck?

At its simplest, a truck is a self-propelled, wheeled vehicle designed to transport goods. All trucks, despite their age, size, and function, consist of a cab or driver's compartment, mounted on a chassis or supporting structure, with wheels. Yet, this does not begin to signal the sheer variety of vehicles, configurations, or indeed the terminology encompassed by the word "truck". It can describe anything from panel vans and pickups, to massive road trains towing multiple trailers. If trucks are diverse, so too are the words used to describe them. Hence, while the unit that pulls a semi-trailer is technically called a "tractor", they are equally known as trucks, "rigs" in the US, and "lorries" in the UK – a play on the archaic verb "lurry", meaning "to pull". While most trucks are classed as light, medium, or heavy, they can range from ultra-light mini-trucks, which can weigh less than 700 kg (1,500 lb), to ultra-heavy off-road haulers that can top 350 tonnes (344.5 tons), not counting their loads.

This book showcases the diversity of trucks over the decades. It includes models designed for personal, commercial, military, and industrial purposes. It even spotlights some especially built for arts and entertainment, from the showman's engine that used to haul the funfair to town, to today's arena-filling monster trucks. The chapters show the many ways that, since their introduction in the 19th century, trucks have helped to shape and build the modern world, and its global consumer society.

The first trucks were steam-powered traction engines that powered farm machinery and pulled trailers. While these mammoth machines could not haul enormous loads by road, they marked a critical step in the development of the purpose-built, self-propelled vehicles that would eventually replace horse-drawn wagons and carts.

Advances in internal combustion engines in the late 19th century paved the way for petrol-powered trucks – the first went on sale in 1896. In their earliest motorized form, trucks consisted of a petrol engine mounted onto a chassis, usually derived from a car (although some also came from former horse-drawn carts), with wheels clad in solid rubber. Although primitive and fragile, they could carry goods faster than a horse, and offered a greater load capacity. Although it would take some years before they could match the power of steam, these pioneering trucks found a market in local delivery operations.

Historically, wartime tends to accelerate technological advances in the years that follow, and the "Roaring 20s" were no exception. Many large truck manufacturers had made their name during World War I, and continued to build on that foundation in the postwar years. Returning servicemen who had learned to drive during the war gave nations a new supply of highly experienced truck drivers, and new vehicles were developed for a recovering world. Inflated pneumatic tyres became the norm, powerful diesel engines were developed, and the modern truck began to take shape

The Great Depression in 1929 bankrupted many founding truck manufacturers in the US, such as Hawkeye, Stewart, and Rugby. However, many others were absorbed into expanding conglomerates that had the combined resources and market share necessary, not only to survive the period, but also to innovate. For example, the cab-over design was pioneered by The White Company in 1932, while that same year, International launched a diesel-powered semi tractor with a 16.3-tonne (16.1-ton) towing capacity, heralding a new age of diesel in the trucking industry.

World War II brought about the greatest demand in truck manufacturing the world had seen. In the lead-up to the conflict, Mercedes-Benz and Opel were forced to abandon civilian truck manufacturing, when they and other leading German motor manufacturers were pressed into the service of the German army, developing new military vehicles. Meanwhile, British manufacturers, such as Bedford, Austin, and Morris, were joined by counterparts in the US, namely Diamond T, GMC, Studebaker, Ford, and Chevrolet, in supplying trucks for the Allied nations. After D-Day at Normandy in 1944, a convoy of 6,000 trucks, many driven by African American troops, formed the "Red Ball" supply lines that helped speed up liberation. This period also saw the emergence of the GAZ truck company in the USSR.

During the postwar years in France, Renault and Berliet resumed making trucks, with Berliet launching the world's largest truck to date in 1957 – the T100. In the US, truck owners picked up on the hot-rod craze sweeping the country, while turbocharged diesel engines led to a new generation of powerful rigs and pickups. Long Australian road trains became possible in the 1960s, as improved engines allowed for bigger loads. Previously, Australia had largely relied upon British trucks, but as American turbo-powered engines surpassed those of their rivals, US trucks started to dominate the market.

Heavy trucks, in particular, entered popular culture in the 1970s, making their big-screen debut and inspiring a growing trucking community that many young people now aspired to join. Across Europe, manufacturers produced vehicles for home and export markets, while in India, Tata had grown from a partnership with Mercedes-Benz to build and supply trucks for Asian markets.

By the 1980s, heavy trucks featured air brakes, air suspension, and power steering as standard, while what would seem like ridiculously powerful engine options became hot-sellers in pickups. Biodiesel, a more eco-friendly fuel, also saw its mainstream introduction decades after its first appearance, and trucks began to be fitted with sophisticated electronic systems that have continued to evolve to offer drivers greater levels of refinement, efficiency, and safety.

Recent advances in truck design have focused on the fuels that power them in a drive for greater sustainability. Little more can be done to advance the aerodynamic efficiency of truck bodies, but a lot can still be achieved with their fuel. Many manufacturers are now prioritizing the development of battery-electric models. Hydrogen fuel cell technology is on the horizon, while the rise of autonomous vehicles suggests a future age of driverless, zero-emissions trucking.

Though trucks have changed immensely in a century and half of developments, their basic function and objective has not: to transport a load or equipment from one location to another swiftly, economically, and safely. In fulfilling this function, trucks have become the backbone of the world's distribution systems, integral not only to transport but to public utilities, waste management, agriculture, forestry, mining, and the construction industries. Most of the goods and services that modern society relies upon have been carried at some point by a truck.

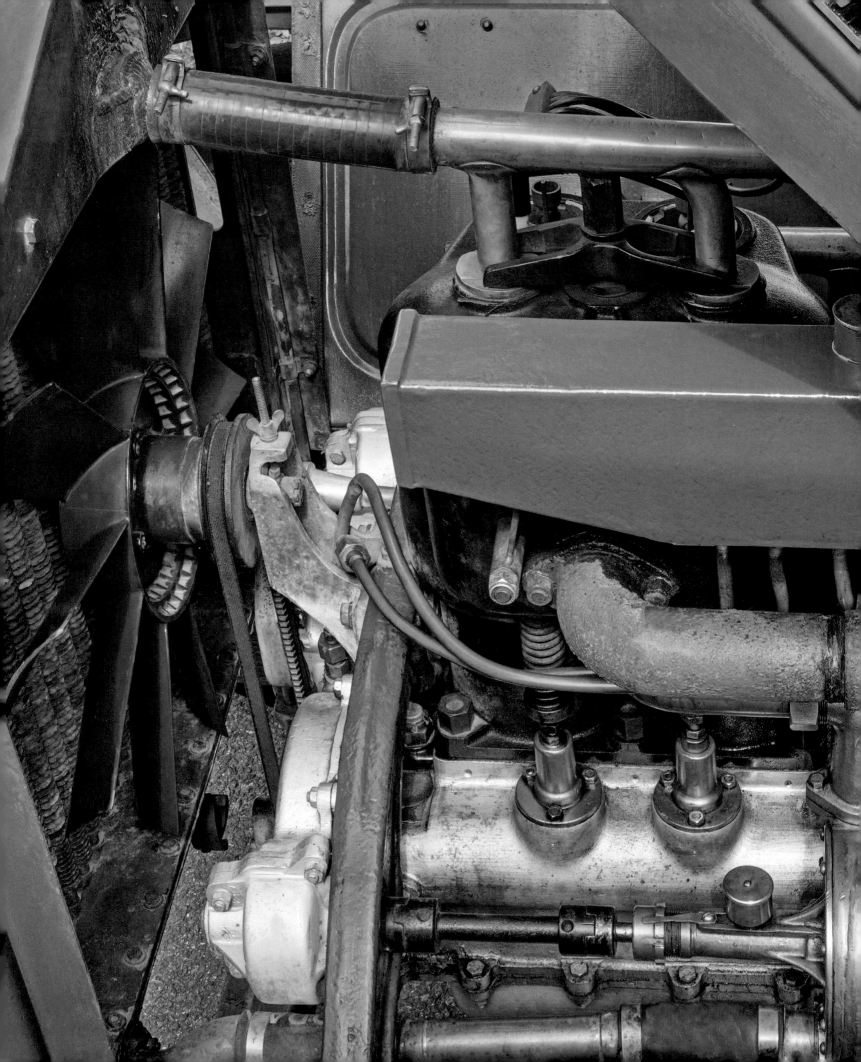

# THE EARLY YEARS

**Animals had carried the burden** of transporting goods by road for centuries, then were joined in the mid-19th century by the steam-powered traction engines developed in the UK. Their time- and cost-saving benefits quickly became apparent, as traction engines offered greater power than horses and more flexibility than steam trains – although their payload was smaller than a train, they were not limited to travelling on fixed rails to a set timetable.

Traction engines were then followed by steam trucks that offered an even more flexible solution for freight. Unlike traction engines that pulled trailers, the steam truck carried its load on the rear of the chassis behind the cab. This basic formula in four-wheeled truck design has changed little over 130 years, even if the means to power it has. The arrival of the 20th century heralded an exciting future that promised to be easier and faster, fuelled by petrol, diesel, or electricity.

By 1914, the truck had played a large part in shaping our increasingly mechanized world. With the onset of World War I, there was a surge in demand, not only for thousands of trucks, but also the people to build and drive them.

△ **Horses and trucks**
Ex-War Department lorries typically shared the roads with horse and carts into the 1930s.

However, the truck and its forerunner, the horse and wagon, continued to serve together.

Following the war, the world looked towards a more positive future with a vibrant, modern outlook. A wide range of petrol and electric trucks with wagon-like wheels, began appearing on the roads alongside horse-drawn carts. However, the introduction of the first diesel-powered vehicle in 1920 changed everything.

## Key events

▷ **1896** Daimler-Benz produces the world's first truck with an internal combustion engine.

△ **Daimler and Benz**
Gottlieb Daimler (right) with Karl Benz and Wilhelm Maybach during the demonstration of the 6 hp Daimler truck at the Paris Motor Show in June 1898.

▷ **1902** Scania in Sweden builds its first truck. Petrol-powered, it carries a payload of 1.5 tonnes (1.5 tons), and can reach just over 11 km/h (7 mph).

▷ **1910** Australian firm Caldwell Vale produces the first large truck to feature 4-wheel drive and power steering.

▷ **1910** Gearbox transmissions start to replace chain and sprocket drive, improving ride and performance.

▷ **1910** The first truck to bear the name Mack appears. The brand becomes synonymous with heavy US trucks.

▷ **1911** Czechoslovakian company Praga builds its first commercial vehicle.

▷ **1914** German company Fruehauf demonstrates its first articulated truck. Both brand and concept become instrumental in global trucking.

▷ **1914** Ford Model TT launches. Longer than previous car-derived pickups and vans, it features enhanced brakes.

▷ **1919** London department store Harrods acquires a fleet of small electrically-powered delivery vans.

▷ **1919** Surplus wartime trucks enter the market, leading to small haulage firms.

> "The **gasoline truck** is gradually creeping up on the **electric** truck and **leaving** the **horse-drawn** truck **far behind**."

*SCIENTIFIC AMERICAN*, 18 JANUARY 1913

◁ **Bethlehem Motor Truck Corporation, USA,** advertisement from 1918 evokes its wartime contributions.

# Steam Trucks

Steam power transformed the emerging industrial world from the early 1800s. French engineer Nicolas-Joseph Cugnot built the first steam-driven road vehicle in 1769, decades ahead of the first steam trains. The Industrial Revolution started in Britain and quickly spread throughout the rest of the world, with machines saving time and costs, while speeding up industry. The birth of the steam traction engine, which was used to pull trailers or power machinery, enabled agricultural advances, such as large-scale ploughing and harvesting with new specialist-designed machines. Traction engines were then joined on the roads by other types of steam truck, which were adopted across many industries.

THORNYCROFT STEAM VAN
THE STEAM CARRIAGE & WAGON Cº CHISWICK

◁ **Thornycroft Steam Van**

**Date** 1896  **Origin** UK

**Engine** 2-cylinder vertical compound

**Payload** 1 tonne (1 ton)

Founded in 1866, Thornycroft was one of the first manufacturers of steam-powered lorries. Its first vehicle was powered by a small steamboat engine that drove the front wheels, while the rear wheels provided the steering.

**Heavy-duty** wooden cart wheels

▷ **Thornycroft Steam Lorry**

**Date** 1902  **Origin** UK

**Engine** 2-cylinder under-chassis vertical boiler

**Payload** 2 tonnes (2 tons)

The first 2-tonne steam lorry to be trialled by the British War Office, this was also the first to be made in larger numbers from 1899. A version of it was also used as a London bus.

**Wooden** rear body

**Wooden wheels** with steel bands

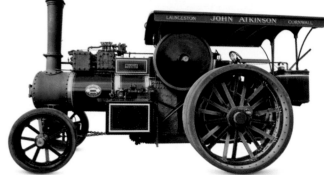

△ **McLaren Road Locomotive**

**Date** 1905  **Origin** UK

**Engine** 2-cylinder

**Payload** Pulling weight of up to 102 tonnes (100 tons)

McLaren's traction engines were manufactured in Yorkshire, UK, and half of its output was exported. Engines like this one were used to haul heavy industrial items, such as boilers and machinery on trailers.

▽ **Bikkers Model C Steam Car**

**Date** 1907  **Origin** Netherlands

**Engine** 1-cylinder vertical boiler

**Payload** 1 tonne (1 ton)

Dutch firm Bikkers built this steam-powered vehicle to tow its maintenance trailer around Amsterdam. The engine powered a pressurized cleaning pump, an early steam cleaner. It had a top speed of 39 km/h (24 mph).

**Roof loading area** for light goods

**Vertical** steering column

**Chimney**

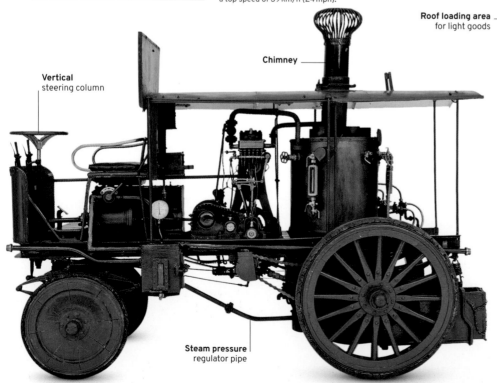

**Steam pressure** regulator pipe

△ **Purrey Steam Truck**

**Date** 1909  **Origin** France

**Engine** 20-bar vertical boiler

**Payload** 5 tonnes (5 tons)

With its wooden, chain-driven wheels wrapped in steel bands, the small Purrey steam truck was used for various light duties around large cities. Some had van bodies and operated at railway stations.

**Cab roof** is a luxury on early steam trucks

△ **Leyland Steam Wagon**

Date 1911    Origin UK

Engine Twin cylinder

Payload 5 tonnes (5 tons)

The Class F version, shown here, was manufactured from 1905–11. This particular wagon was used at Preston docks in Lancashire, not far from where it was built.

**Chimney extension** used when the engine is powering fairground rides

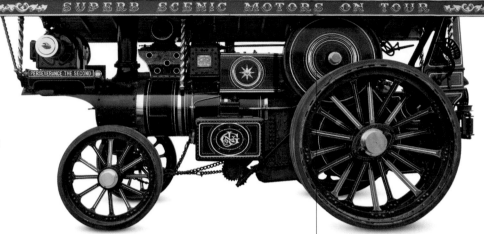

◁ **Burrell Showman's Locomotive**

Date 1913    Origin UK

Engine 2-cylinder

Payload Pulling weights of up to 82 tonnes (81 tons)

The richly decorated traction engine was used to haul the fair and its crew from location to location. Once there, the engine was kept running on idle to provide power for many of the fairground rides.

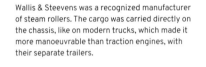

**Large flywheel** connects to the generator at the front via a belt

**Timber frame** and clad cab

◁ **Wallis & Steevens Overtype 5-ton Steam Lorry**

Date 1912    Origin UK

Engine 2-cylinder

Payload 5 tonnes (5 tons)

Wallis & Steevens was a recognized manufacturer of steam rollers. The cargo was carried directly on the chassis, like on modern trucks, which made it more manoeuvrable than traction engines, with their separate trailers.

**Water tank** beneath chassis

◁ **Mann's Steam Wagon**

Date 1920    Origin UK

Engine 2-cylinder

Payload 3 tonnes (3 tons)

This wagon, produced by Yorkshire-based company Mann, had a top speed of 19 km/h (12 mph). Mann manufactured its first steam trucks in 1898 and they were popular with local authorities and contractors alike.

**Boiler inspection cover** used for maintenance

**Rubber road tyres** cut down on noise

**Boiler fire box** and ash pan at the bottom

# The New Fuels

This period marked the dawn of the internal combustion engine patented for motor vehicle use in 1886, and with it the arrival of motor cars. Although there was a wide range of vehicle types available, with electric vehicles driving alongside steam and petrol, only one of these would win through. Steam trucks had to carry coal and water, which made them very heavy with limited range, and electric at that time could not provide either the speed or range needed. The petrol engine was smaller and lighter, and needed less fuel than a steam truck, giving it greater range.

### ▽ Daimler Lastwagen

| Date 1896 | Origin Germany |
|---|---|
| Engine | 1-cylinder 4 hp petrol |
| Payload | 1.5 tonnes (1.5 tons) |

Daimler's first truck was built around a converted horse-drawn wagon, with the 1060cc engine driving the rear wheels via a belt. Like a steam traction engine, the truck used a chain system for steering and wooden blocks pressed against the wheels for brakes.

### △ Daimler 6hp Shooting Brake

| Date 1897 | Origin UK |
|---|---|
| Engine | 2-cylinder petrol engine, 6 hp |
| Payload | 450 kg (992 lb) |

In 1896, a British consortium acquired the rights from Daimler to build and sell its own vehicles in the UK. The first British Daimler established a distinctive bonnet shape. This vehicle was used up until 1924.

**Wooden** cargo bed

**Leaf springs** similar to those used on horse-drawn carts

### ▽ Daimler Lastwagen

| Date 1898 | Origin Germany |
|---|---|
| Engine | 2- and 4-cylinder petrol engines, up to 10 hp |
| Payload | 1.2–5 tonnes (1.2–5 tons) |

This was the first German Daimler truck to feature its engine in front of the driver. The later 1899 model could be fitted with a range of engines depending on its intended use. German brewery companies were among the first to operate these trucks.

**Engines** are located at the back, powering the rear wheels

**Driving seat** similar to that on a horse-pulled wagon

### △ Tatra

| Date 1899 | Origin Austria-Hungary |
|---|---|
| Engine | Two 2-cylinder petrol, 13.2 hp |
| Payload | 2.5 tonnes (2.5 tons) |

Czech firm Tatra started producing vehicles as early as 1850 and unveiled its first truck in 1898. It used two Benz engines, although it could function on just one depending upon how much weight it was carrying.

**Chain** drives the rear wheels

**Crank handle** for starting the engine

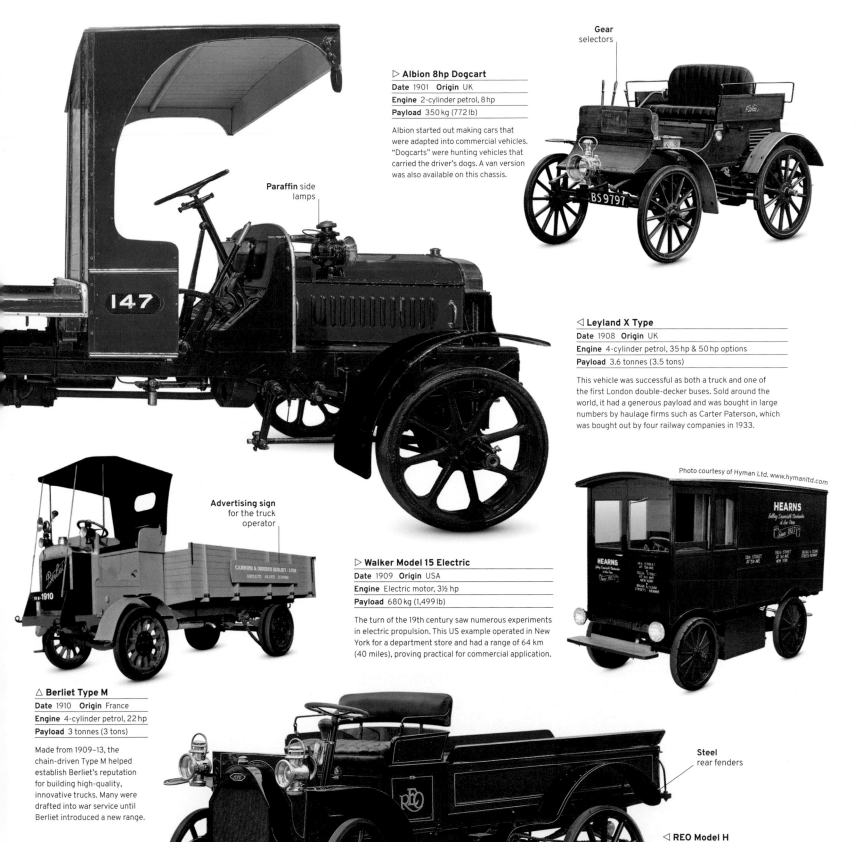

Gear selectors

**▷ Albion 8hp Dogcart**

**Date** 1901  **Origin** UK

**Engine** 2-cylinder petrol, 8 hp

**Payload** 350 kg (772 lb)

Albion started out making cars that were adapted into commercial vehicles. "Dogcarts" were hunting vehicles that carried the driver's dogs. A van version was also available on this chassis.

**Paraffin** side lamps

147

**◁ Leyland X Type**

**Date** 1908  **Origin** UK

**Engine** 4-cylinder petrol, 35 hp & 50 hp options

**Payload** 3.6 tonnes (3.5 tons)

This vehicle was successful as both a truck and one of the first London double-decker buses. Sold around the world, it had a generous payload and was bought in large numbers by haulage firms such as Carter Paterson, which was bought out by four railway companies in 1933.

Photo courtesy of Hyman Ltd. www.hymanltd.com

HEARNS

**Advertising sign** for the truck operator

**▷ Walker Model 15 Electric**

**Date** 1909  **Origin** USA

**Engine** Electric motor, 3½ hp

**Payload** 680 kg (1,499 lb)

The turn of the 19th century saw numerous experiments in electric propulsion. This US example operated in New York for a department store and had a range of 64 km (40 miles), proving practical for commercial application.

**△ Berliet Type M**

**Date** 1910  **Origin** France

**Engine** 4-cylinder petrol, 22 hp

**Payload** 3 tonnes (3 tons)

Made from 1909–13, the chain-driven Type M helped establish Berliet's reputation for building high-quality, innovative trucks. Many were drafted into war service until Berliet introduced a new range.

**Steel** rear fenders

**Engine** is mounted on its side under the driving seat

**◁ REO Model H**

**Date** 1911  **Origin** Canada

**Engine** 1-cylinder petrol, 9 hp

**Payload** 680 kg (1,499 lb)

The first REO trucks were built in Ontario, Canada. The delicate-looking Model H could carry a hefty payload. It also had a twin chain-drive – most other trucks then had just one – and it sold well.

**Chain drive** – one on each side

# Leyland Steam Wagon

Founded in 1896, Leyland was one of the earliest producers of steam wagons. These successors to the horse-drawn cart transported anything and everything in the first decades of the 20th century. The company was one of the few to move away from steam, building its first internal combustion-powered London bus in 1905. By the mid 1930s, however, production of steam-propelled road vehicles in the UK had almost ceased.

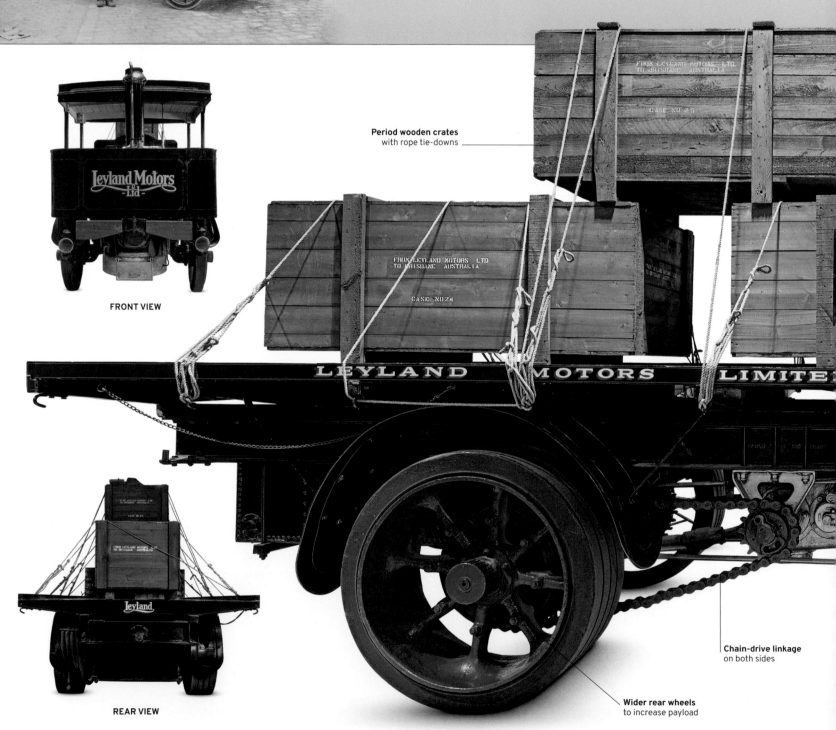

**FRONT VIEW**

**Period wooden crates**
with rope tie-downs

**REAR VIEW**

**Chain-drive linkage**
on both sides

**Wider rear wheels**
to increase payload

## SPECIFICATIONS

| Model | Leyland Steam Wagon |
|---|---|
| Origin | UK |
| Assembly | Lancashire, UK |
| Production run | Unknown |
| Weight | 9.1 tonnes (9 tons) |
| Payload | 5.1 tonnes (5 tons) |
| Engine | 60 bhp |
| Transmission | Chain-drive 2-speed gearbox |
| Maximum speed | 16 km/h (10 mph) |

LEYLAND PRODUCED ITS FIRST steam wagon in 1896 and its expertise with steam propulsion led to the introduction of the Model H in 1905. This new wagon rivalled those of other manufacturers, such as Yorkshire, Robey, and Sentinel. At this time, internal combustion engines could not match the power offered by steam, which could haul much heavier loads, often with a trailer. Many of these steam wagons excelled in dockyards, transporting large quantities of coal and freight quickly to customers. Wagons such as the Model H remained in use well into the 1930s.

**Back from the outback**

This steam wagon was originally exported to Australia and survived long enough to be discovered in the 1970s in the outback, where it had spent its life transporting timber around New South Wales. After it was rescued, it was shipped back to the Leyland factory in the UK, where apprentices conducted a full restoration.

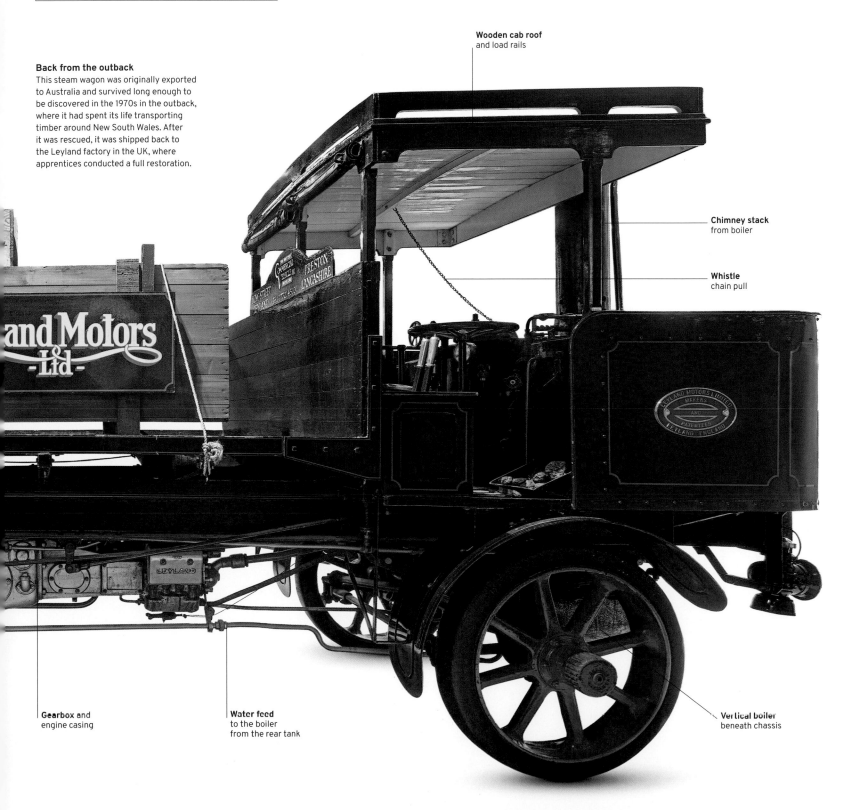

**Wooden cab roof** and load rails

**Chimney stack** from boiler

**Whistle** chain pull

**Gearbox** and engine casing

**Water feed** to the boiler from the rear tank

**Vertical boiler** beneath chassis

## THE EXTERIOR

Although still in their infancy when it was built, the solid rubber tyres of the Leyland Model H help reduce noise and vibration within the vehicle. The bodywork is minimal: at the front is a slab-sided cab to enclose the driving controls and the boiler; at the back is a flat, wooden loading bed to carry the goods. The water tank is located under the chassis at the rear, from where it can be easily replenished.

The design sets out the standard layout that trucks still use today, with a cab at the front and a long loading bed stretching out behind. Steam wagons proved to be a clever combination of traction engine and truck. They were the first step in the evolution of powered road haulage.

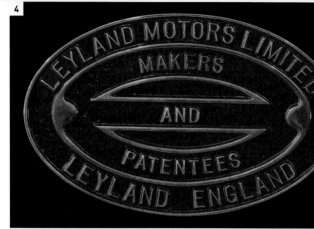

**1.** Acetylene lamps  **2.** Towing hitch  **3.** Front wheel showing detachable hub  **4.** Brass-maker's plate

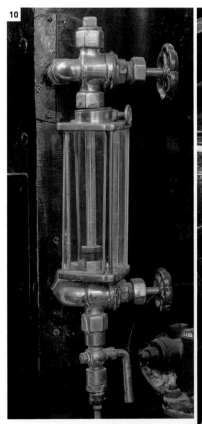

## THE ENGINE AND GEARBOX

Placed under the chassis to help maximize space for loads, the vertical boiler has a working pressure of 1,400 kPa (200 psi). Daily tasks include emptying the ash pan and cleaning the boiler, while checking for leaks, and topping up water and coal. While traditional traction engines offer easy accessibility, the Model H's underslung engine means operators have to crawl on their hands and knees to carry out maintenance.

**5.** Steam expansion chamber  **6.** Gearbox with twin chain-drive sprockets  **7.** Gearbox and final drive sprocket to axle  **8.** Gearbox casing  **9.** Engine cylinders

## THE INTERIOR

Driving steam wagons is challenging and demands great attention. Everything in the cab area is extremely hot – much welcomed on cold, frosty days but less so in the height of a busy summer. The brakes are operated by winding the lever until the brake blocks meet the wheels.

Steam engines work on the principle of building up pressure, which is then released into the cylinders to provide power, as and when it is needed. One of the most important features on any steam engine is the pressure release valve. Hence the only two gauges in this wagon are the large brass pressure gauge and the glass sight gauge for monitoring water levels. The vertical steering column has a direct link to the front axle via a worm drive, rather than a chain link, which traction engines use. Having a roof over the cab in 1905 would have been a luxury.

**10.** Sight glass water-level gauge  **11.** Driving controls, vertical steering column  **12.** Coal bunker  **13.** Engine drive controls and gear selection

# Trucks for the Masses

As businesses saw the advantages of commercial vehicles, demand grew significantly, requiring enormous production runs. With lessons learned and experiences gained from wartime production, factories set about improving models for new global civilian businesses. Henry Ford revolutionized vehicle production in 1913 with the introduction of the world's first moving assembly line, which sped up production and saved money. Manufacturing techniques – such as the use of steam presses to cut out parts and hot-riveting to join them – were adopted from other industries, such as shipbuilding and boilermaking. Tools were also improved at this time.

THE SOUTHERN COUNTIES AGRICUL
TRADING SOCIETY LTD..
CORN & SEED MERCHANTS
WINCHESTER.
TELEGRAMS, FARMERS.

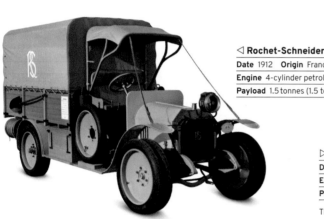

◁ **Rochet-Schneider**

| | | | |
|---|---|---|---|
| Date | 1912 | Origin | France |
| Engine | 4-cylinder petrol, 12 hp | | |
| Payload | 1.5 tonnes (1.5 tons) | | |

Better known for building early racing cars and expensive grand tourers for nobility, Rochet-Schneider supplied small, high-endurance trucks like this for expeditions into the Sahara.

▷ **Ford Model T Mail Truck**

| | | | |
|---|---|---|---|
| Date | 1913 | Origin | USA |
| Engine | 4-cylinder petrol, 20 hp | | |
| Payload | 340 kg (720 lb) | | |

The open body of this mail van made it easy to access the packages. To produce greater traction, a zinc compound was added to tyre moulds, which turned them white in the process.

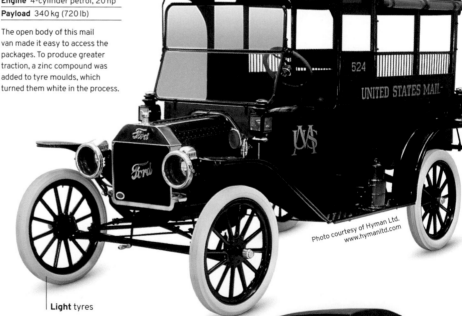

524

UNITED STATES MAIL

Photo courtesy of Hyman Ltd.
www.hymanltd.com

**Light** tyres

## TECHNOLOGY

## Uses of Truck Chassis

It was commonplace for car chassis to be fitted with different commercial bodies to cater to different trades. But because load capacities were restricted to under a tonne in most cases, larger purpose-built truck chassis were also used. Even so, the payloads of heavier trucks were still limited by the available power of their engines, with most payloads in the 3–5 tonne (3–5 ton) range during this period. The same chassis could be used as a bus, military vehicle, fire engine, or delivery truck.

**Dennis Fire Engine** This 1908 engine is typical of those used by factories' in-house fire departments.

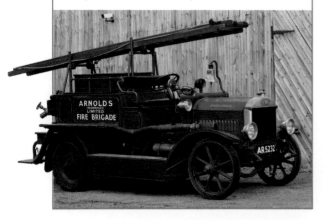

ARNOLDS LIMITED FIRE BRIGADE
AR 5232

▷ **Hawkeye Model K**

| | | | |
|---|---|---|---|
| Date | 1916 | Origin | USA |
| Engine | 4-cylinder petrol, 30 hp | | |
| Payload | 1.5 tonnes (1.5 tons) | | |

Hawkeye made this first truck out of its factory in Iowa. Engines were bought in from Buda, a firm with long experience in engine manufacturing. Hawkeye trucks were sold as far afield as China and the UK.

**Wooden** cargo body

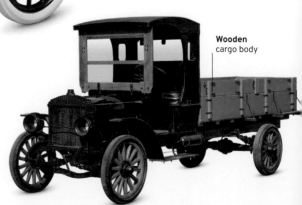

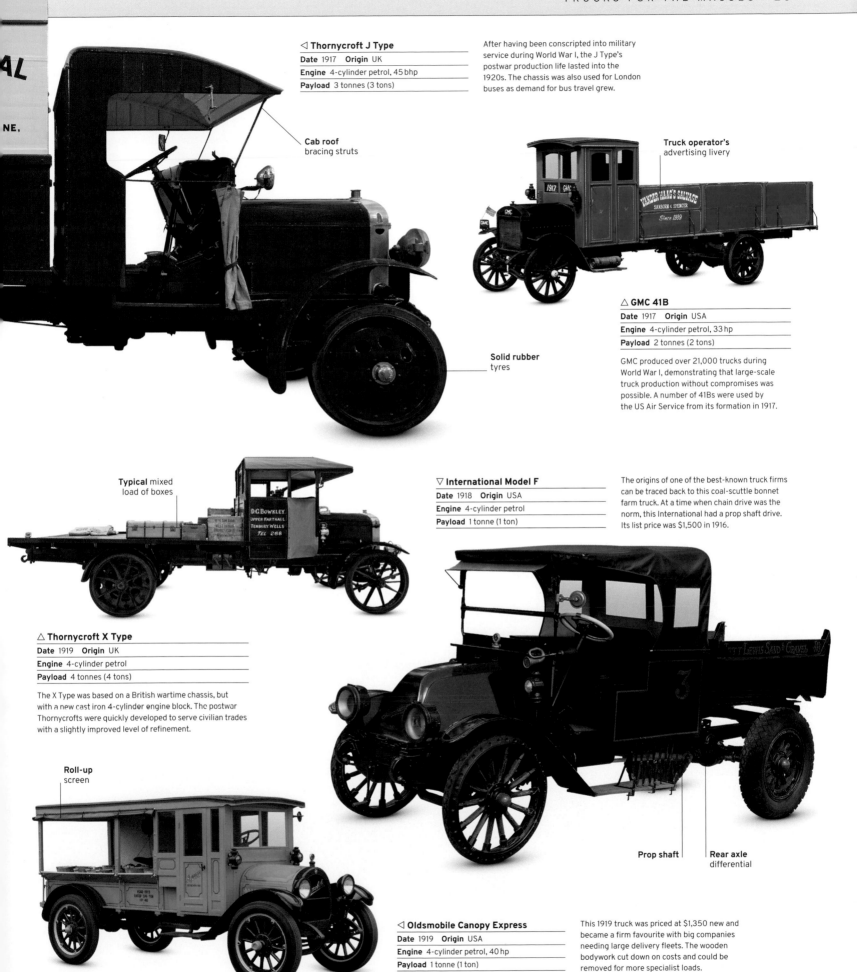

◁ **Thornycroft J Type**

**Date** 1917 **Origin** UK

**Engine** 4-cylinder petrol, 45 bhp

**Payload** 3 tonnes (3 tons)

After having been conscripted into military service during World War I, the J Type's postwar production life lasted into the 1920s. The chassis was also used for London buses as demand for bus travel grew.

Cab roof
bracing struts

Truck operator's
advertising livery

Solid rubber
tyres

△ **GMC 41B**

**Date** 1917 **Origin** USA

**Engine** 4-cylinder petrol, 33 hp

**Payload** 2 tonnes (2 tons)

GMC produced over 21,000 trucks during World War I, demonstrating that large-scale truck production without compromises was possible. A number of 41Bs were used by the US Air Service from its formation in 1917.

Typical mixed
load of boxes

△ **Thornycroft X Type**

**Date** 1919 **Origin** UK

**Engine** 4-cylinder petrol

**Payload** 4 tonnes (4 tons)

The X Type was based on a British wartime chassis, but with a new cast iron 4-cylinder engine block. The postwar Thornycrofts were quickly developed to serve civilian trades with a slightly improved level of refinement.

▽ **International Model F**

**Date** 1918 **Origin** USA

**Engine** 4-cylinder petrol

**Payload** 1 tonne (1 ton)

The origins of one of the best-known truck firms can be traced back to this coal-scuttle bonnet farm truck. At a time when chain drive was the norm, this International had a prop shaft drive. Its list price was $1,500 in 1916.

Prop shaft

Rear axle
differential

Roll-up
screen

◁ **Oldsmobile Canopy Express**

**Date** 1919 **Origin** USA

**Engine** 4-cylinder petrol, 40 hp

**Payload** 1 tonne (1 ton)

This 1919 truck was priced at $1,350 new and became a firm favourite with big companies needing large delivery fleets. The wooden bodywork cut down on costs and could be removed for more specialist loads.

# Women, Trucks, and the Road to War

The United States entered World War I in 1917, and life changed radically for women there. While some ran their husbands' businesses, others worked as drivers and mechanics, and for the first time entered the world of heavy manufacturing.

## SUPPLIES FOR SOLDIERS

Grace Gallatin Seton Thompson, a leader of the Connecticut Women's Suffrage Association in the 1910s, founded Le Bien-Être du Blessé (Welfare of the Injured) Women's Motor Unit in 1916. This all-female service was set up to send trucks and supplies directly to soldiers on the Western Front.

Taken in Central Park in 1917, not far from Grace's home in Greenwich, this image shows two Ford Model T trucks that would go on to see action in Europe, moving people and goods within the war zone. The message on the door of each truck reads: "Women's City Club of New York. This truck the gift of Allied Festa, Utica, N.Y."

**Two members of Le Bien-Être du Blessé** stand beside vehicles that will soon be sent to Europe.

# Key Manufacturers
# The Leyland Story

Leyland is one of the most famous truck brands in the UK. The manufacturer's history goes back more than 120 years, and in that time, it has produced some innovative and iconic trucks. While the Leyland badge is not seen on the road anymore, the company is still going strong, manufacturing trucks for DAF, another famous marque.

**Leyland steam-powered truck, 1919**

**LEYLAND'S ORIGINS DATE** back to the early days of motorized transport. In 1896, Henry Spurrier and James Sumner founded The Lancashire Steam Motor Company to produce a steam-powered van they had built. The van, which had a capacity of 1.5 tonnes (1.5 tons), soon proved to be a success. Buoyed by this, the company produced more vehicles, including buses, and created the prototype for its first petrol-engine vehicle in 1904, affectionately nicknamed "The Pig".

In 1907, the company acquired steam wagon builder Coulthards of Preston and was renamed Leyland Motors Company. It also expanded its North Works factory in Lancashire, England, to keep up with demand.

In 1912, as the prospect of war loomed, Leyland moved into the military market with its 3-tonne truck, usually referred to as the RAF Type. During WWI , Leyland built nearly 6,000 vehicles for the British Forces. While the early

**An early Leyland badge**

1920s were tough times economically for Leyland, the company survived and returned to prosperity later in the decade with a new range of buses and trucks. Many had whimsical animal names, such as the Terrier or the Llama, and quickly became customer favourites. The Leyland "Zoo" names would be synonymous with the manufacturer for several decades. In the 1930s, trucks such as the Hippo, Rhino, Octopus, and Buffalo were added to the heavy range of vehicles, and diesel engines were also introduced.

Post WWI saw great expansion of the Leyland operation, including the acquisition of car and commercial vehicle manufacturer Albion Motors in 1951 and truck marque Scammell in 1955. Trucks were produced under the Albion and Scammell brands until 1972 and 1988 respectively. The company also exported trucks globally, including to India, where they were badged Ashok Leyland. Leyland continued to innovate during the 1950s. Its all-steel Vista Vue cab,

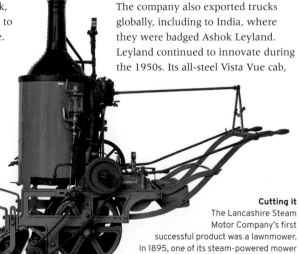

**Cutting it**
The Lancashire Steam Motor Company's first successful product was a lawnmower. In 1895, one of its steam-powered mower engines was fitted to a three-wheel car.

**Merryweather fire engine, 1922**
Leyland began in 1896 by producing steam-powered vehicles, such as the above fire engine. Steam models were phased out in favour of internal combustion engines by 1926.

with enhanced visibility, debuted in 1958. It was replaced six years later by the Ergomatic cab, which offered new levels of driver comfort and safety. However, its most prominent feature was that it could be tilted forward, giving better access to the engine than had been possible with previous fixed cab designs. The cab was so popular that slightly updated versions of it were still being fitted to Leyland trucks in 1981.

Leyland expanded its operations in the 1960s. It acquired Associated Commercial Vehicles in 1962, which included marques such as AEC and Thornycroft. Car and aero engine maker The Rover Company Limited was purchased in 1966, followed by Aveling-Barford in 1967, which made road rollers and dumper trucks. In 1968, Leyland Motors merged

with British Motor Holdings, which included truck maker Guy, to form British Leyland Motor Corporation (BLMC), but this merger was not a success. Given the number of companies that were part of the group – many of which made similar products – BLMC was difficult to manage, leading to financial problems. In 1974, it received a guarantee from the British Government, but this was not enough, and in 1975 the company went bankrupt and was nationalized as British Leyland. The company was split into four divisions, including

**"The Pig"**

1896 Formation of The Lancashire Steam Motor Company
1897 First van wins top prize at Royal Agricultural Society of England trials
1904 First petrol-fuelled vehicle, "The Pig" is developed
1905 Production of first petrol-powered vehicle, a double-decker bus
1907 Acquisition of Coulthards of Preston; renamed Leyland Motors Company
1912 Introduction of RAF Type 3-tonne truck for military market

**Albion Claymore**

1951 Albion Motors is acquired
1955 Leyland acquires Scammell, another truck manufacturer
1958 Introduction of Vista Vue cab
1962 Acquisition of Associated Commercial Vehicles, including marques AEC and Thornycroft
1964 Introduction of Ergomatic tilting cab
1966 Acquisition of car and aero engine manufacturer Rover
1967 Acquisition of road roller and dumper truck manufacturer Aveling-Barford

**Red Line 420 FG**

1968 Leyland Motors merges with British Motor Holdings to form British Leyland Motor Corporation (BLMC)
1975 BLMC goes bankrupt and is nationalized as British Leyland, including Leyland Truck & Bus
1980 Launch of T45 series, with GWV range of 10–39 tonnes (9.8–37.4 tons)
1981 Leyland Trucks and Leyland Bus split into separate divisions
1981 T45 series wins International Truck of the Year Award

**DAF XG Plus**

1987 Leyland Trucks merges with Dutch rival DAF Trucks to create DAF NV; Leyland-DAF brand is used in UK
1993 DAF goes bankrupt; UK truck division bought by the management and renamed Leyland Trucks
1998 Leyland Trucks is bought by US company PACCAR
2000 Leyland-DAF name is retired – trucks now solely sold as DAF
2021 The 500,000th truck rolls off the Leyland production line, a DAF LF 210

Leyland Truck & Bus, which operated within the Land Rover Leyland Group. This division was then split into two – Leyland Trucks and Leyland Bus – in 1981.

Leyland Trucks had something of a resurgence in this period, with the launch of the T45 range. Developed with Ogle Design, the series included the Roadtrain, Roadrunner, Freighter, Cruiser, and Constructor models. The T45 scooped the International Truck of the Year Award for 1981.

Despite the award, sales declined in the UK and abroad. In 1987 this led to Leyland's merger with Dutch rival DAF, forming DAF NV. However, sales continued to fall, and in 1993 DAF NV went bankrupt. Following a management buy-out, the UK truck division was renamed Leyland

Trucks, while a new DAF company was reformed in the Netherlands. Soon after, an arrangement with DAF was established that saw Leyland selling DAF trucks to the UK market as Leyland-DAF.

In 1998, Leyland Trucks was bought by PACCAR, the US company that also bought DAF. In 2000, the Leyland name was dropped from trucks, but this did not mean the end of the company. Its Lancashire factory still employs 1,000 workers producing 12,000 DAF trucks a year.

" Whatever the need, **wherever the need**, under the most rigorous conditions, you can **rely on Leyland** for the toughest of hauls!"

LEYLAND TRUCKS ADVERTISEMENT, 1948

**Leyland DAF DROPS, 2010**
The sturdy, off-road Leyland DAF DROPS (demountable rack offload and pickup system) with a 15-tonne (14.8 ton) payload have been in service with the British Army since 1990.

# Light Delivery Vans

With the arrival of small cars powered by internal combustion engines, local distribution networks began running on petrol as well as steam. Small business owners increasingly used light delivery vans to handle around-town freight. These early commercial vehicles were very simple at first – a box body attached to a chassis would suffice. The small car chassis and patented engines of European automotive manufacturers De Dion-Bouton and Benz helped pioneer the modern small trademan's vehicle.

▽ **Benz No.1: Velo**

**Date** 1894  **Origin** Germany
**Engine** Single-cylinder petrol, 3 hp
**Payload** 150 kg (330 lb)

One of the first vehicles with an internal combustion engine, this model became very popular in Europe. It consisted of a car chassis with a wooden box for cargo, tiller steering, and chain drive.

**Brake** lever

**Front** scuttle panel

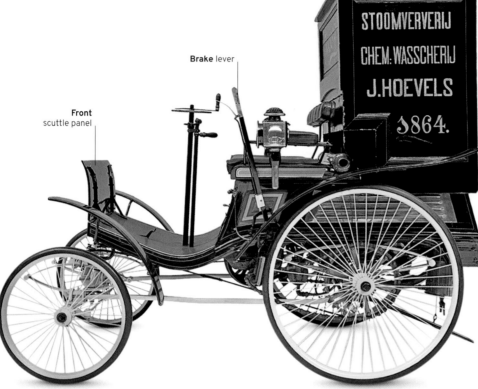

▽ **Ford Steam Bread Van**

**Date** 1902  **Origin** UK
**Engine** Unknown
**Payload** 250 kg (550 lb)

During this period, existing horse-drawn vehicles were often converted to steam or petrol power. Although the example shown was built in 1970, it uses a 1902 van body and petrol engine, and is typical of such a conversion.

**Tiller-operated** steering

Photo courtesy of Hyman Ltd.
www.hymanltd.com

△ **Oldsmobile Model R**

**Date** 1904  **Origin** USA
**Engine** Single-cylinder, 4.5 hp
**Payload** 150 kg (330 lb)

First made in 1901, this "Curved Dash" Oldsmobile – so called for the shape of its dashboard – was one of early motoring's biggest success stories. It was offered as a delivery van by various coach builders.

▽ **Cadillac Model A6**

**Date** 1904  **Origin** USA
**Engine** Petrol, 6.5 hp
**Payload** 250 kg (550 lb)

Better known for their luxurious motorcars, Cadillac offered various made-to-order vans up until the 1940s. Cadillac's first commercial vehicle was a Model A car adapted with a factory-made van body.

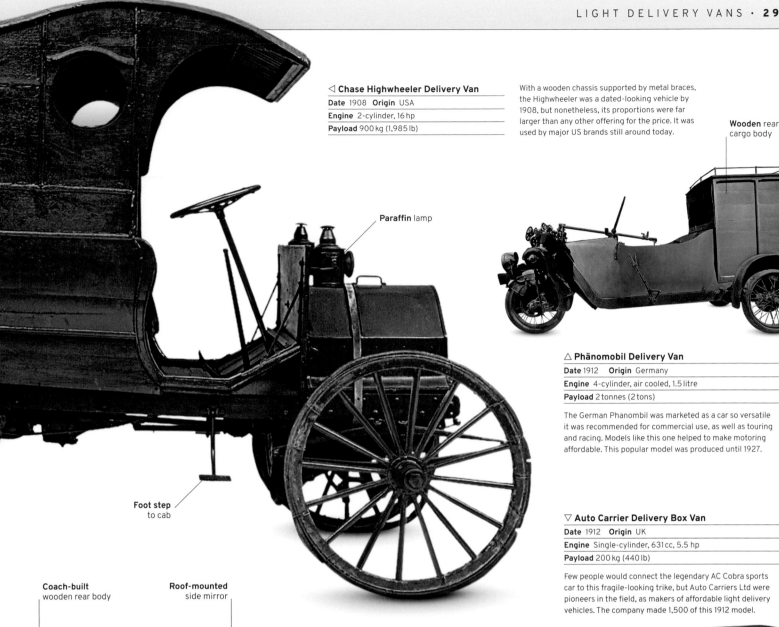

◁ **Chase Highwheeler Delivery Van**

| Date 1908 | Origin USA |
|---|---|
| **Engine** 2-cylinder, 16 hp | |
| **Payload** 900 kg (1,985 lb) | |

With a wooden chassis supported by metal braces, the Highwheeler was a dated-looking vehicle by 1908, but nonetheless, its proportions were far larger than any other offering for the price. It was used by major US brands still around today.

**Wooden** rear cargo body

**Paraffin** lamp

**Foot step** to cab

△ **Phänomobil Delivery Van**

| Date 1912 | Origin Germany |
|---|---|
| **Engine** 4-cylinder, air cooled, 1.5 litre | |
| **Payload** 2 tonnes (2 tons) | |

The German Phanombil was marketed as a car so versatile it was recommended for commercial use, as well as touring and racing. Models like this one helped to make motoring affordable. This popular model was produced until 1927.

▽ **Auto Carrier Delivery Box Van**

| Date 1912 | Origin UK |
|---|---|
| **Engine** Single-cylinder, 631 cc, 5.5 hp | |
| **Payload** 200 kg (440 lb) | |

Few people would connect the legendary AC Cobra sports car to this fragile-looking trike, but Auto Carriers Ltd were pioneers in the field, as makers of affordable light delivery vehicles. The company made 1,500 of this 1912 model.

**Coach-built** wooden rear body

**Roof-mounted** side mirror

◁ **Fiat 18P Tipo**

| Date 1915 | Origin Italy |
|---|---|
| **Engine** 4-cylinder petrol, 4.4 litres | |
| **Payload** 2 tonnes (2 tons) | |

Fiat vehicles sold all over Europe and particularly well in the UK, where they were used as taxis and vans. Large orders were placed during WWI for Fiat vans to be used as transport trucks and field ambulances. This commercial model used the Fiat truck chassis that was developed for military vehicles.

# Pierce-Arrow R Type

Loved by presidents and movie stars alike, Pierce-Arrow was one of the leading US luxury motorcar manufacturers between 1901 and 1938 – based at their plant in Buffalo, New York. When World War I broke out in 1914, the company, like many others, switched production from cars to trucks, and other vehicles, to support their armed forces during the conflict. Pierce-Arrow produced many such trucks for the Allied forces.

**THE R8 5-TON TRUCK** was one of many Pierce-Arrow vehicles delivered to the French government in 1916. The journey from the US to France for these trucks was perilous: vehicles were hoisted onto barges at the docks in New York and towed across the Atlantic to Le Havre, but many did not survive the rough sea passage.

After the war, the truck pictured here was retained in military service until the late 1920s, when it was sold for civilian purposes in the south of France. It was brought to the UK in 1991 and restored. It is shown here in British Army paintwork with a 13-pounder anti-aircraft gun mounted in the back.

13lb shells

Brass kerosene side lamp

Radiator filling cap

Radiator design inspired by Dennis lorries engineer

Front mudguard

Solid rubber tyre

Bonnet side panel easily removed

Tool box compartment

Tigress

8675

LOAD EXCEED

## SPECIFICATIONS

| | |
|---|---|
| Model | Pierce-Arrow R Type |
| Origin | USA |
| Assembly | Buffalo, New York |
| Production run | 1911–19, approx. 16,500 |
| Weight | 4.5 tonnes (4.5 tons) |
| Payload | 3 tonnes (3 tons) |
| Engine | 4-cylinder petrol, 3880 cc, 30 hp |
| Transmission | 3-speed manual |
| Maximum speed | 29 km/h (18 mph) |

**FRONT VIEW**

**CLOSED REAR VIEW**

LOAD NOT TO EXCEED 3 TONS

↑ 18675

SV·4680

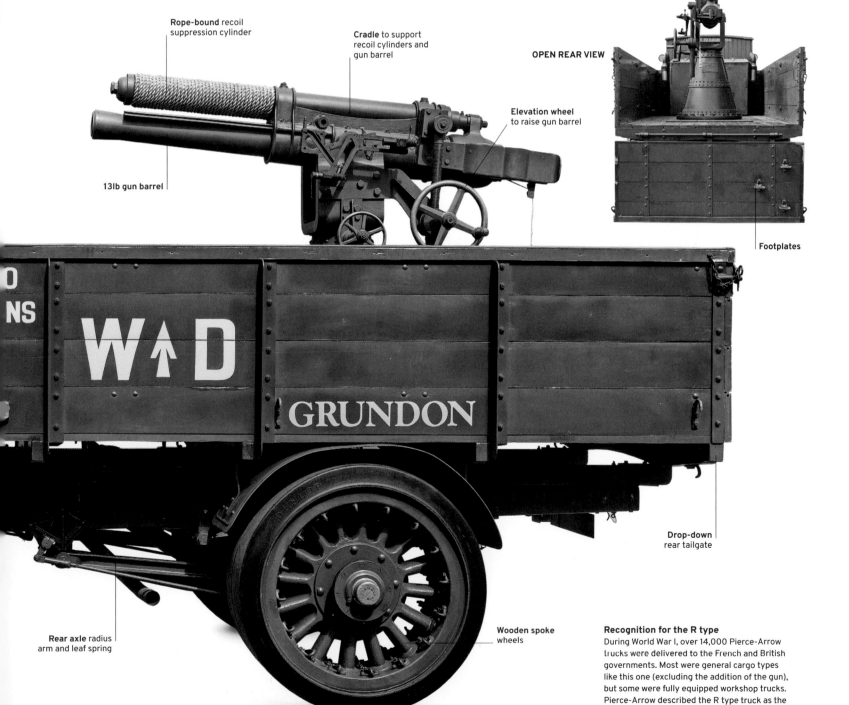

**OPEN REAR VIEW**

Rope-bound recoil suppression cylinder

Cradle to support recoil cylinders and gun barrel

Elevation wheel to raise gun barrel

13lb gun barrel

Footplates

Drop-down rear tailgate

Rear axle radius arm and leaf spring

Wooden spoke wheels

### Recognition for the R type

During World War I, over 14,000 Pierce-Arrow trucks were delivered to the French and British governments. Most were general cargo types like this one (excluding the addition of the gun), but some were fully equipped workshop trucks. Pierce-Arrow described the R type truck as the "American Rolls-Royce of trucks".

## THE EXTERIOR

Painted a drab green camouflage colour scheme, this truck has no windshield or doors. Travelling in it would have left a driver exposed to the elements, although some trucks were fitted with a canvas apron to offer a measure of warmth and wet-weather protection for the crew. Glass was forbidden due to the risk of shell bursts causing shattering. With the exception of the bonnet and wings, the bodywork is mostly wood.

**1.** Maker's badge  **2.** Chassis plate with identification numbers  **3.** Main headlamp, fitted for wartime use  **4.** Fuel feed from tank selector lever  **5.** War Department vehicle number and adopted name  **6.** Heavy-duty wooden spoked wheel with detachable rim  **7.** Bonnet catch  **8.** Rear spring hanger bracket  **9.** Footplates to gain access

## THE INTERIOR

Easy access is central to the design of the cab. There is little to get in the way to impede rapid entry and exit – essential in a wartime situation. The foot pedals are typical of the era, with a centre throttle pedal, brake pedal on the right, and clutch pedal on the left. When driving at night during the war, only one light (if any) was used, and that would be the one closest to the driver, so that it could be extinguished quickly.

## THE ENGINE

Brass was often used for the external metal parts of various engine elements – the water pump (seen bottom centre in image 17) being one such example. Having a pump at all was quite a sophisticated feature for a truck engine of the period. The four-cylinder engine comprises two blocks, each housing two cylinders, that bolt on to a common alloy crankcase. The engine is started by turning a handle that protrudes through the front of the vehicle.

17. Engine bay; the bonnet hinges upwards to provide access to the engine
18. Engine bay, showing the magneto – an early form of ignition system

10. Dashboard and driving controls
11. Steering wheel and advance/retard ignition control levers   12. Sight glass for fuel tank level   13. Ignition switch and engine starter button   14. Cab-mounted fire extinguisher   15. Foot pedals: the centre pedal is the accelerator   16. Handbrake lever and gear lever selector

A Model T truck being used to load hay on a farm in England c.1930

# Key Manufacturers
# The Ford Story

A forerunner in its field, the Ford name has been key to the development of both cars and trucks for well over 100 years. While now synonymous with pickup trucks in the US, Ford also has a long history of producing heavier trucks, which continues to this day in Europe.

HENRY FORD ALWAYS had a fascination with how things worked. He left his family farm in Michigan at 16 to apprentice in a machine shop, eventually working out how to make his first car – the Quadricycle – in 1896 with a four-hp, two-cylinder engine.

In 1903, Ford and his investors founded the Ford Motor Company with $28,000 in capital. His first car, the Model T, was introduced in 1908 at a price of $825. The company sold 15 million of them before production ceased. The first purpose-built Ford

**Henry Ford**
(1863–1947)

truck was the Model TT, which had a 0.9-tonne (1 US ton) capacity. It was based on the Model T car but had a longer wheelbase and heavier frame. The TT remained in production until 1928, when it was replaced by the Model A, which – like its predecessor – was designed to be both affordable and dependable. Featuring the flathead V8 engine, still beloved by today's car enthusiasts, the Model A also proved popular and remained in production for over 22 years. Ford's successful F-series line of trucks was then launched in 1948. Featuring a

heavier-duty platform, they were known as "Bonus Built" trucks. The range was available in eight size and weight ratings, with capacities from 0.5–3 tonnes (0.5–3 tons). The 1948 F-1 was known as the "million dollar cab", as that was the amount Ford spent redesigning the cab for the series.

In the 1960s, Ford streamlined its heavy truck range. In 1961, the heavy-duty F-750 and F-1100 became

a separate model line, along with a new H-series Linehauler. The range was added to in 1963 with the medium-duty N-series Super Duty, which shared a grille with the H-series and a cab with the F-series

**Special delivery**
The Ford F-5, launched in 1948, was the first truck to have its chassis developed specifically for truck use. It also marked Ford's entry into the medium truck segment.

> "Ford is credited with putting the **world on wheels**, and Ford trucks helped."
>
> BOB KREIPKE, FORD HISTORIAN

**Mass production**
Truck chassis fade into the distance on this conveyor, part of the assembly line in Chicago, 1925.

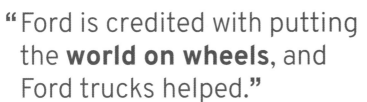

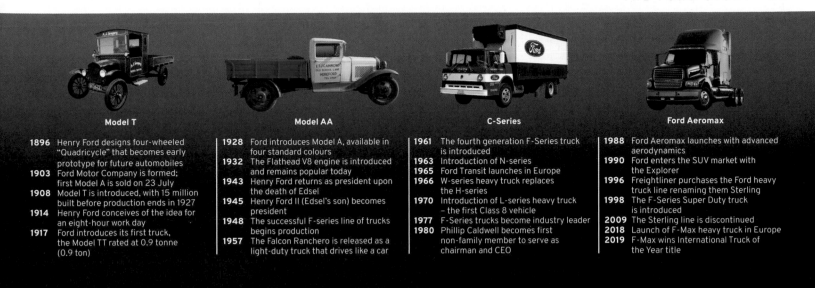

**Model T**

**Model AA**

**C-Series**

**Ford Aeromax**

| | | | |
|---|---|---|---|
| **1896** Henry Ford designs four-wheeled "Quadricycle" that becomes early prototype for future automobiles | **1928** Ford introduces Model A, available in four standard colours | **1961** The fourth generation F-Series truck is introduced | **1988** Ford Aeromax launches with advanced aerodynamics |
| **1903** Ford Motor Company is formed; first Model A is sold on 23 July | **1932** The Flathead V8 engine is introduced and remains popular today | **1963** Introduction of N-series | **1990** Ford enters the SUV market with the Explorer |
| **1908** Model T is introduced, with 15 million built before production ends in 1927 | **1943** Henry Ford returns as president upon the death of Edsel | **1965** Ford Transit launches in Europe | **1996** Freightliner purchases the Ford heavy truck line renaming them Sterling |
| **1914** Henry Ford conceives of the idea for an eight-hour work day | **1945** Henry Ford II (Edsel's son) becomes president | **1966** W-series heavy truck replaces the H-series | **1998** The F-Series Super Duty truck is introduced |
| **1917** Ford introduces its first truck, the Model TT rated at 0.9 tonne (0.9 ton) | **1948** The successful F-series line of trucks begins production | **1970** Introduction of L-series heavy truck – the first Class 8 vehicle | **2009** The Sterling line is discontinued |
| | **1957** The Falcon Ranchero is released as a light-duty truck that drives like a car | **1977** F-Series trucks become industry leader | **2018** Launch of F-Max heavy truck in Europe |
| | | **1980** Phillip Caldwell becomes first non-family member to serve as chairman and CEO | **2019** F-Max wins International Truck of the Year title |

pickup trucks. In 1966, the W-series was introduced as a replacement for the H-series. Meanwhile, in Europe, Ford had launched its Transit van, which became an instant success. It went on to become Europe's bestselling van for four decades. Ford then introduced the L-series heavy-duty truck in 1970. It was their first vehicle to be called a Class 8, according to new US Department of Transportation guidance. Because they were manufactured in Louisville, Kentucky, they were often called the "Ford Louisville". In 1988, Ford launched its Aeromax tractor unit – an upgrade of the existing L-9000. The new truck had many enhancements and was one of the most aerodynamic on the market at the time. It also had – a first for a truck in North America – car-like style composite headlights. In the following years, other trucks were added to the Aeromax line, including commercial and urban delivery trucks, as well as in 1992, the LLA and LTLA-9000, which had extended hoods and set-back front axles.

The L-series, as well as all Louisville, Aeromax, and Cargo lines, were sold to Daimler-owned Freightliner in 1996. Renamed the Sterling Line, the trucks were produced in Ontario, Canada, from 1998 until 2009, when Freightliner shut down Sterling operations.

While Ford has exited the heavy commercial truck market in the US and South America, it is still active in Europe through its Ford Otosan subsidiary, based in Türkiye. In 2018, Ford Otosan launched the F-Max, which scooped the 2019 International Truck of the Year title. Ever popular in the US, Ford trucks are growing in popularity in central and southern Europe, as well as in the Middle East.

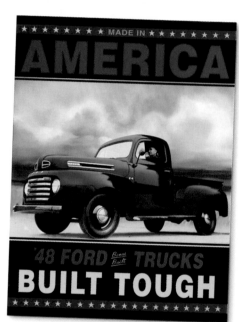

**Advertisement for Ford trucks, 1948**
Ford invested heavily in its trucks in the postwar period. In 1948, trucks had standard features including a driver's side sun visor, which was unusual at the time.

**Heavy duty**
Ford exited the US heavy truck market in the mid-1990s, selling its lines to Freightliner. It is still active in the heavy truck market in Europe, however.

# World War I Trucks

This period was the age of mechanical transport and modernized warfare. As war broke out in 1914, huge numbers of trucks were needed to work alongside horses and there simply was not enough time to develop military trucks in the quantity needed. To fill the gap, civilian vehicles were drafted in, hastily painted green and deployed to the front. AEC London buses were used as troop carriers and pigeon lofts for signals. Driving along, these vehicles were terrifyingly vulnerable and steering proved heavy in the mud. The lessons learned from how the trucks coped in extreme conditions paved the way for a new generation of vehicles that could negotiate tough terrain, while the hardy drivers, shaken and now highly experienced, had the skills to take on new roles when they returned home.

Bulkhead and engine firewall

△ **Fiat 15 Ter**

| Date 1915 | Origin Italy |
|---|---|
| **Engine** 4-cylinder petrol, 35 hp | |
| **Payload** 1.5 tonnes (1.5 tons) | |

Italy's first military truck was the medium-sized Fiat 15 Ter. Large numbers of this 1.5-tonne light truck were also used by the British, French, Russian, and US armies throughout the war.

△ **Latil TAR**

| Date 1915 | Origin France |
|---|---|
| **Engine** 4-cylinder petrol, 30 hp | |
| **Payload** 4 tonnes (4 tons) | |

Three thousand of these 4x4 Latil trucks were built during the war for pulling heavy artillery. They had a 20.3 tonne (20-ton) towing capacity. They were one of the first to replace horses in this role. Many were still in service at the start of World War II.

Paraffin-powered headlamp

△ **Pierce Arrow R8**

| Date 1916 | Origin USA |
|---|---|
| **Engine** 4-cylinder petrol, 38 hp | |
| **Payload** 3.6 tonnes (3.5 tons) | |

During the war, Pierce Arrow built over 14,000 R8 trucks, with their cross-braced chassis design. A team of factory engineers accompanied each overseas shipment to get the trucks ready.

Wooden cargo body

△ **Berliet CBA**

| Date 1916 | Origin France |
|---|---|
| **Engine** 4-cylinder petrol, 25 hp | |
| **Payload** 3.5 tonnes (3.5 tons) | |

The 6-tonne Berliet CBA became the most used French truck during Word War I. Berliet built more than 25,000 of this iconic model for the Allied forces. Many remained in service until World War II.

△ **Thornycroft J Type**

| Date 1916 | Origin UK |
|---|---|
| **Engine** 4-cylinder petrol, 40 hp | |
| **Payload** 3 tonnes (3 tons) | |

The bulk of the British forces' army trucks consisted of 3-tonne general-purpose cargo types, such as the J Type, of which more than 5,000 were built. Many were shipped to India and Egypt, and lasted for decades longer than expected.

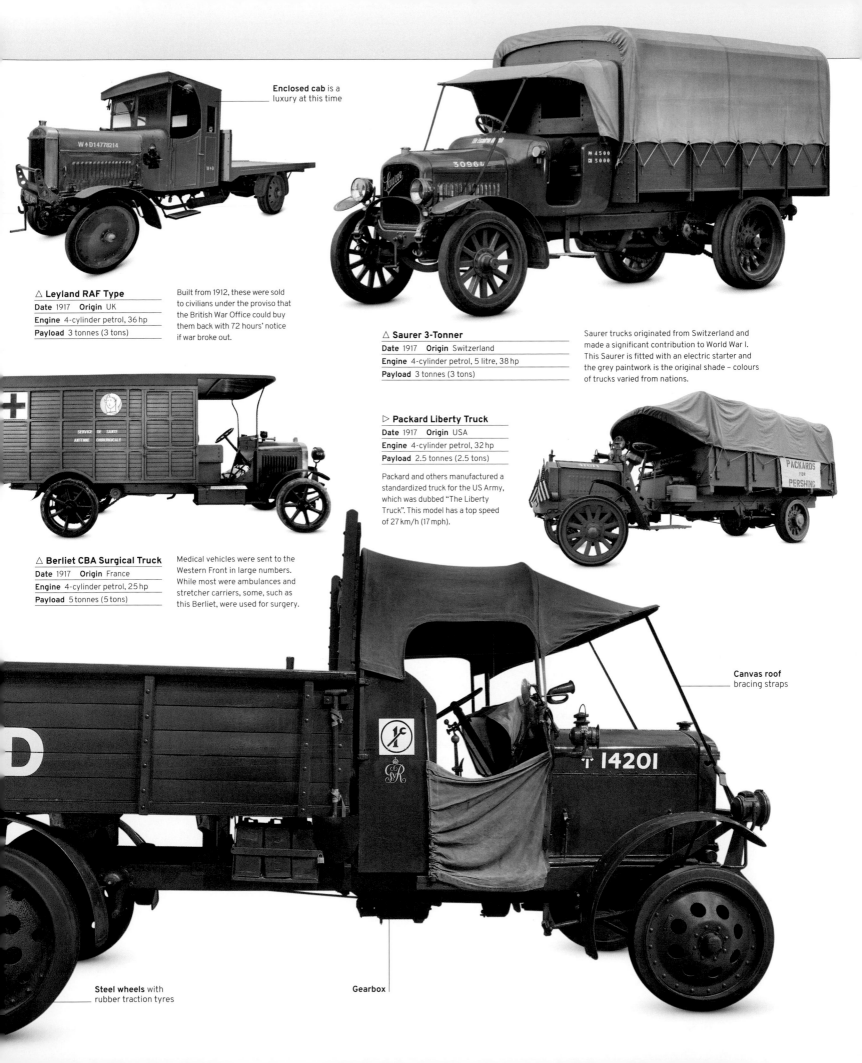

**Enclosed cab** is a luxury at this time

### △ Leyland RAF Type

Date 1917   Origin UK

Engine 4-cylinder petrol, 36 hp

Payload 3 tonnes (3 tons)

Built from 1912, these were sold to civilians under the proviso that the British War Office could buy them back with 72 hours' notice if war broke out.

### △ Saurer 3-Tonner

Date 1917   Origin Switzerland

Engine 4-cylinder petrol, 5 litre, 38 hp

Payload 3 tonnes (3 tons)

Saurer trucks originated from Switzerland and made a significant contribution to World War I. This Saurer is fitted with an electric starter and the grey paintwork is the original shade – colours of trucks varied from nations.

### ▷ Packard Liberty Truck

Date 1917   Origin USA

Engine 4-cylinder petrol, 32 hp

Payload 2.5 tonnes (2.5 tons)

Packard and others manufactured a standardized truck for the US Army, which was dubbed "The Liberty Truck". This model has a top speed of 27 km/h (17 mph).

### △ Berliet CBA Surgical Truck

Date 1917   Origin France

Engine 4-cylinder petrol, 25 hp

Payload 5 tonnes (5 tons)

Medical vehicles were sent to the Western Front in large numbers. While most were ambulances and stretcher carriers, some, such as this Berliet, were used for surgery.

**Canvas roof** bracing straps

**Steel wheels** with rubber traction tyres

**Gearbox**

# Many Hands Make Heavy Work

Long before automated production lines, every part of making and assembling a truck was carried out by skilled mechanics. Large or small, each job required a human touch. Before Ford's introduction of the assembly line in 1913, truck manufacturers built each vehicle one at a time and in one place, often using differing (rather than standardized) parts. The process took more than a day's work instead of a few hours, and the craftspeople involved required precise skills. The jobs they did corresponded to their expertise, and every job – from painting the body to assembling the gearbox – was done by hand using simple tools.

## MANPOWER AND HORSEPOWER

In the early days of the Swedish company Scania, their artisans made bicycles, cars, and trucks out of the same Mälmo factory. This photo shows the 40-plus mechanics required to assemble the early Scania truck pictured and its 24-hp engine, meaning there is more manpower than horsepower on show ("manpower" because women would not enter into heavy manufacturing until the following decade).

**A 4-tonne Scania Type E flatbed model** from 1908, carrying the entire team that produced it.

# Powering Ahead

In the 1890s, German engineer Rudolph Diesel devised a new, more energy-efficient type of internal combustion engine. Both the engine and the fuel it uses now bear his name, and they have powered trucks for nearly a century. However, other power sources and fuels were also viable options for a time.

**Cart and truck, London, 1922**
In the early decades of the 20th century, horse-drawn vehicles and motorized trucks worked side by side on city streets. Here, a brewer's cart, called a dray, is seen beside an electric truck operated by the same brewery.

Until the end of the 19th century, the only "trucks" transporting goods by road were horse-drawn wagons and carts, with "engines" (the horses) fuelled by hay and oats. For decades, horse-drawn vehicles coexisted with the early mechanized trucks, and held their own. Horses were cheap and plentiful, and more suited to the frequent stop-starts of delivery rounds than internal combustion engines or steam-powered vehicles, and could negotiate poor-quality roads and rural tracks better than trucks were able to.

## Steam's brief moment

It was not always obvious that petrol and diesel would be the prime power providers for trucks. One early rival was steam, which had been used for decades. At the beginning of the 20th century, steam-powered trucks were a common sight in the US and Europe.

After World War I, demand for steam power declined as the internal combustion engine began to prove its superiority – petrol trucks were cheaper and more efficient. Steam's fate was sealed by the combination of more restrictive legislation on road steam haulage, and the increasing cost of buying and running them. A few examples remained on UK roads, however, until the 1960s.

## Alternative options

Trucks with electric motors powered by batteries were also popular in the first truck decades, especially in cities, where they did not need to drive as far. They fell out of favour as roads improved and other types of vehicles could achieve a far greater range. Walker Electric Trucks was making battery vehicles in the US as early as 1907, and by 1913 GMC was offering a full line of trucks powered by "Edison current", including heavy-duty models.

During World War I, some civilian trucks were converted to run on gas derived from coal – or even wood. The gas was stored in a huge bag secured to the roof of the truck.

In 1923, American firm Autocar launched its E1 electric truck, which could haul up to 1 US ton (0.9 tonnes/0.9 tons). The E3 and

**Rudolph Diesel (1858–1913)**
Diesel tested a number of fuels in his engine, including petrol. The most successful eventually proved to a fuel called distillate – a by-product of oil refining that was often discarded. This fuel is now known as diesel.

## Alternative Fuels

With environmental concerns on the rise, and strict climate targets, the days of the diesel engine are numbered. As a result, a range of alternatives are in development, but most mean investment in new trucks, which can be expensive. However, hydrotreated vegetable oil (HVO) provides a great option that can reduce carbon emissions by up to 90 per cent without the need to buy new trucks. HVO is made from animal fats or vegetable oils, which are processed using "hydrotreatment" at high temperature and pressure to produce a stable fuel with a long life. It has many of the same chemical properties as fossil diesel, which is why it is sometimes referred to as "green diesel". HVO can be used in standard truck diesel tanks without any detrimental effect to the engine or performance and has lower emissions than diesel.

**This Volvo FE tanker is carrying HVO**, a fossil-free diesel made by hydrogenating plant-based oils and waste. It is less environmentally harmful than conventional diesel.

E5 followed, the numbers indicating their haulage capacity. Commonplace in the 1920s and 1930s, these trucks were much faster than horse-drawn wagons, but cost about the same to buy and run. They had three sets of rechargeable batteries mounted in cradles, which could be swapped in and out when a battery ran flat.

## Fossil fuel wins

In the long run, petrol and diesel trucks offered a far greater range before the need for refuelling than their electric counterparts could. A contributory factor in the decline of electric trucks was lack of charging infrastructure. Battery-electric power survived for specialist uses, including forklift trucks and, especially in the UK, milk floats that delivered to the doorstep.

> " a **steam wagon** travelling... in the **40–50 mph** bracket... was **particularly impressive.** "
>
> PAT KENNETT, *THE FODEN STORY*

The first trucks powered by internal combustion engines burned petrol – a logical development, since petrol had proved itself a successful fuel in early cars that became the basis for early trucks. The wind of change began to blow in 1923, when Karl Benz fitted a diesel engine into a truck. Compared with the performance of a petrol model of a similar design, Benz's diesel truck achieved fuel savings of 86 per cent. In addition to being more efficient, diesel was better suited to trucks than petrol because it is a more powerful fuel. Diesel gives a higher torque output (the turning force produced by the engine), which trucks need in order to carry or haul heavy loads, even when moving at the slowest speeds. During the 1930s, the diesel engine became a mainstream option for trucks in Europe, although in North America, petrol engines still largely held sway.

After World War II, diesel continued its rise to dominance. Petrol remained popular in North America for heavy trucks until the 1970s (and it is still used for some smaller trucks today), but even there diesel eventually won out. Pivotal to this was the full-scale adoption of diesel engines by major truck manufacturers, such as Ford and Mercedes-Benz, confirming diesel as the fuel of choice for trucks.

**Harrods electric van, 1939**
Having used American Walker electric vans since 1919, Harrod's department store in London began building delivery vans in its own workshops in the 1930s. Sixty vans powered by underfloor batteries were built up to 1939.

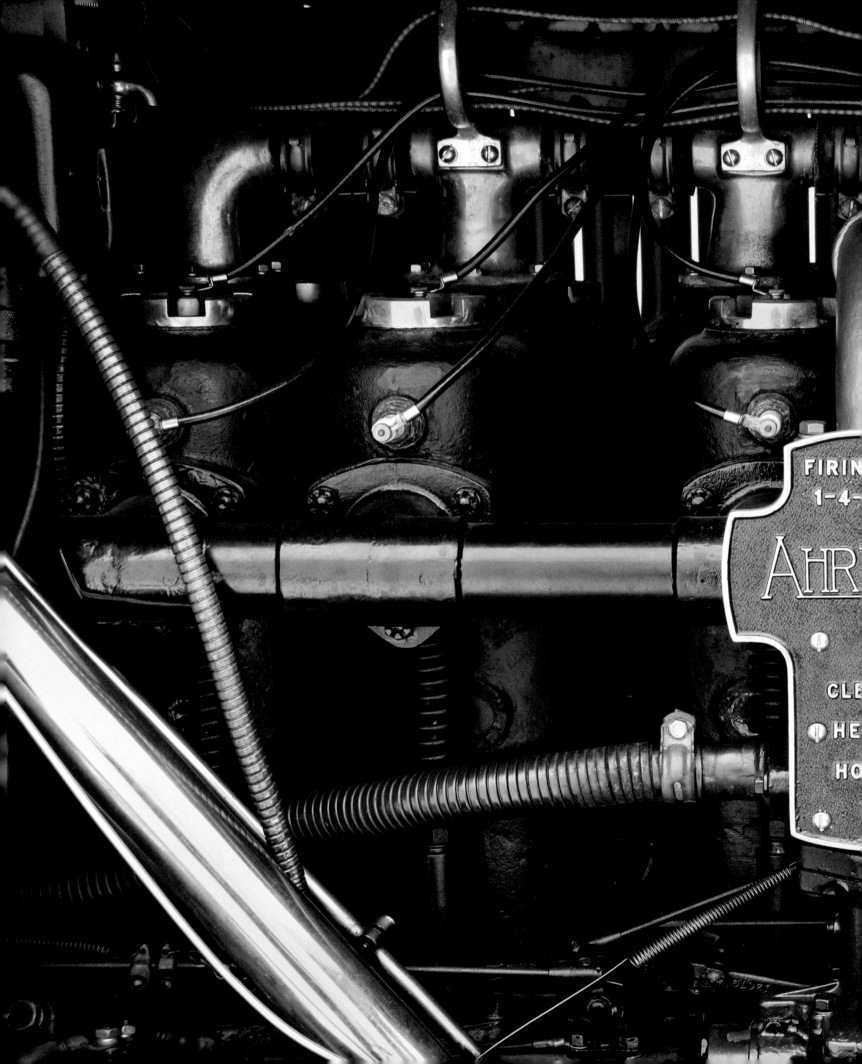

1920–1938
# STYLE AND SUBSTANCE

# STYLE AND SUBSTANCE

**From teapots to express trains,** nothing was left untouched by the artistic trends of the 1920s and 30s. Trucks now featured chrome, comfortable seats, and smooth lines. They were still mainly powered by petrol, but the advances in diesel engines at this time led to engines that powered trucks for the rest of the century. As truck power increased, so did the loads they could haul, allowing them to compete against the railways that had dominated cargo transport up until then.

△ **Art movement**
In the 1930s, trucks took on a more streamlined, Art Deco style that continued into the 1940s. This beer delivery truck has the smooth lines and curves typical of the time.

To many, this era was the golden age of Hollywood, and that glitz and glamour filtered into transport. Trucks representing firms featured elaborate advertising, and the 1930s saw the introduction of streamlined tankers, vans, and goods vehicles inspired by the Art Deco movement. Split-screen windshields added to this modern look, and plastic coverings wrapped around previously bare metal steering wheels and handles to improve safety and comfort.

As trucks developed throughout the 1930s, drivers tested their skills in inhospitable and remote locations that were now more accessible, from the frozen tundra of the Arctic Circle to the Sahara desert. Road infrastructure was expanding with building programmes including German autobahns and US freeways. This connected cities, enabling communities and industries to grow. A distinct "trucking culture" began to evolve amongst truckers and a sense of brotherhood developed on the road.

In 1920, there were roughly **1,108,000 commercial vehicles** on the road in the US. By 1939, this had **increased** to around **4,783,000.**

## Key events

▷ **1921** Scammell is founded in the UK. Its trucks quickly build a solid reputation for haulage and military use.

▷ **1923** Kenworth trucks is established in the US to meet the demands of the timber industry.

▷ **1926** Route 66 is commissioned in the US, connecting Chicago to California. Nicknamed "The Mother Road", it enables trucks to drive coast to coast.

▷ **1927** Ford Model AA is introduced as the replacement for the Model T. It is the first Ford produced in the UK.

▷ **1927** Volvo produces its first truck for the harsh conditions of the Scandinavian climate.

▷ **1929** The Great Depression hits. In the US, many people turn cars into pickup trucks because fuel coupons are given out to truck owners.

△ **Entire life in a truck**
The Great Depression of the 1930s led to sights such as this farmer piling all his belongings into a truck.

▷ **1932** The first Russian GAZ AA truck rolls off the production line – a licensed Ford Model AA.

▷ **1933** American Trucking Association is formed to represent the welfare and needs of truck drivers.

# Heavy Trucks of the 1920s

A key change in the trucks of the "Roaring Twenties" was the adoption of pneumatic (air-filled) tyres, which replaced the solid, bone-shaking rubber ones of previous years. New engines with a single cylinder block simplified production and kept costs down for buyers. Most heavy-trucks over 3 tonnes (3 tons) were fairly rudimentary, with speeds rarely above 32 km/h (20 mph). The poor driver, who had until now been left out in the cold, was shown more sympathy, and glazed cabs soon became the norm. Steam power was still available – and very much in use – but it slowly began to give way to petrol.

◁ **Mack AC "Bulldog"**

| | |
|---|---|
| **Date** 1923 | **Origin** USA |
| **Engine** 4-cylinder petrol, 74 hp | |
| **Payload** 4.5 tonnes (4.5 tons) | |

This chain-drive truck established Mack's reputation for rugged heavy vehicles. Production began in 1915; despite its dated appearance, the "Bulldog" stayed in production (with updates) until 1938, with nearly 40,000 units leaving the factory.

◁ **Renault MZ**

| | |
|---|---|
| **Date** 1925 | **Origin** France |
| **Engine** 4-cylinder petrol, 17 hp | |
| **Payload** 5 tonnes (5 tons) | |

The MZ, Renault's first truck with front brakes, bore the company's typical coal-scuttle bonnet design, and with the radiator behind the engine. The example shown here is a tractor unit for hauling small trailers, and was used to transport ice for fishing boats in Lisbon, Portugal.

◁ **White Model 45 Dumper Truck**

| | |
|---|---|
| **Date** 1925 | **Origin** USA |
| **Engine** 4-cylinder petrol, 45 hp | |
| **Payload** 5 tonnes (5 tons) | |

The semi-open "C" cab design was intended for use on building sites. As well as enabling quick and easy access, it gave good visibility for on-site tasks, such as shifting rubble and construction materials.

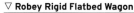

**TECHNOLOGY**

## The First 100-tonne Truck

The 102-tonne (100-ton) Scammell was the first internal combustion truck built to carry loads of up 100 tonnes. Powered by an 80 hp Rolls Royce military engine with double chain drive, it carried a variety of loads – from ship rudders, bridges, and boilers, to bulky items that would not fit on trains – that had previously been moved by steam traction. Such was the enormous size of the vehicle that it required a crew of three to operate it. Its development was mirrored elsewhere: trucks across the world grew in size as the need for heavy haulage increased.

**The huge Scammell 100-tonner** was made of riveted steel plates built into a box frame with apertures for the engine, transmission, and turntable (fifth wheel).

▽ **Robey Rigid Flatbed Wagon**

| | |
|---|---|
| **Date** 1926 | **Origin** UK |
| **Engine** 2-cylinder steam | |
| **Payload** 6 tonnes (6 tons) | |

Beginning life as an articulated tanker, this steam wagon was later converted into a rigid flatbed, making it more versatile and easier to manoeuvre. With coal still plentiful and cheap, steam power remained popular throughout the 1920s.

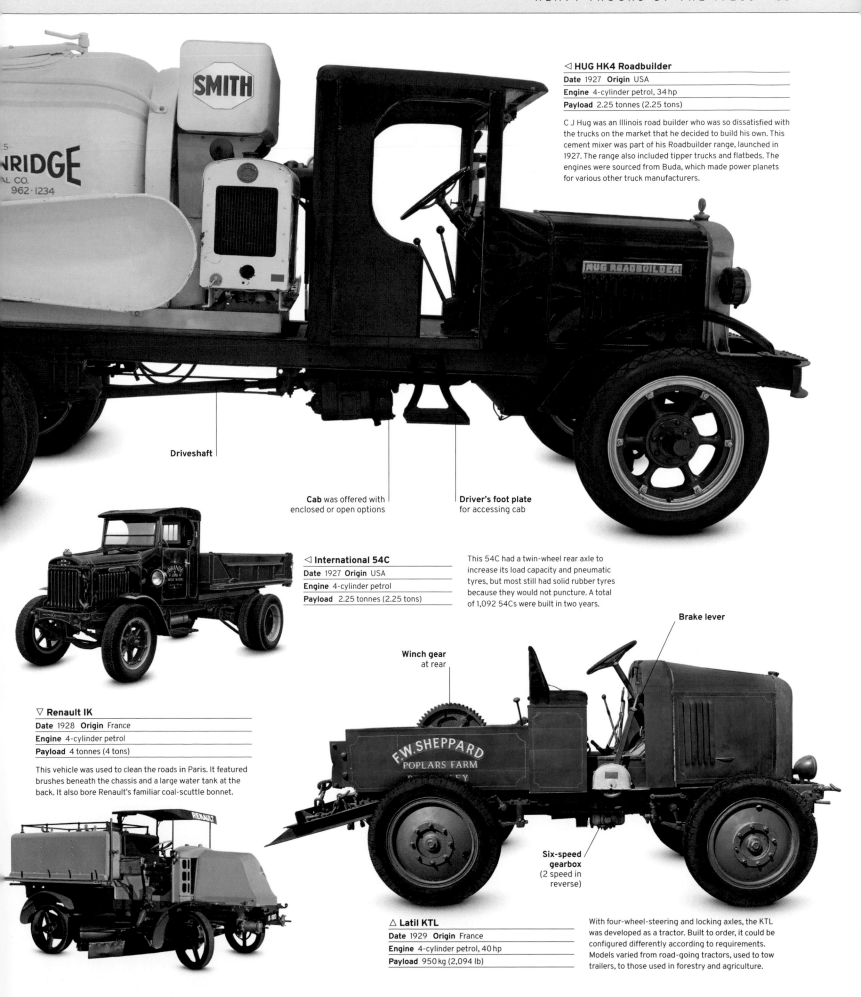

**SMITH**

NRIDGE
AL CO.
962·1234

Driveshaft

**Cab** was offered with
enclosed or open options

**Driver's foot plate**
for accessing cab

◁ **HUG HK4 Roadbuilder**

Date 1927  Origin USA

Engine 4-cylinder petrol, 34 hp

Payload 2.25 tonnes (2.25 tons)

C J Hug was an Illinois road builder who was so dissatisfied with
the trucks on the market that he decided to build his own. This
cement mixer was part of his Roadbuilder range, launched in
1927. The range also included tipper trucks and flatbeds. The
engines were sourced from Buda, which made power planets
for various other truck manufacturers.

◁ **International 54C**

Date 1927 Origin USA

Engine 4-cylinder petrol

Payload 2.25 tonnes (2.25 tons)

This 54C had a twin-wheel rear axle to
increase its load capacity and pneumatic
tyres, but most still had solid rubber tyres
because they would not puncture. A total
of 1,092 54Cs were built in two years.

▽ **Renault IK**

Date 1928 Origin France

Engine 4-cylinder petrol

Payload 4 tonnes (4 tons)

This vehicle was used to clean the roads in Paris. It featured
brushes beneath the chassis and a large water tank at the
back. It also bore Renault's familiar coal-scuttle bonnet.

**Winch gear**
at rear

**Brake lever**

F.W.SHEPPARD
POPLARS FARM

**Six-speed
gearbox**
(2 speed in
reverse)

△ **Latil KTL**

Date 1929 Origin France

Engine 4-cylinder petrol, 40 hp

Payload 950 kg (2,094 lb)

With four-wheel-steering and locking axles, the KTL
was developed as a tractor. Built to order, it could be
configured differently according to requirements.
Models varied from road-going tractors, used to tow
trailers, to those used in forestry and agriculture.

# Key Manufacturers
# The Renault Trucks Story

French manufacturer Renault is one of the leading European names in trucks today, but its complicated history involves mergers with and acquisitions of many different manufacturers over the years. The company's reputation for producing innovative trucks can be traced back more than a century to two motoring pioneers.

**Berliet Type L Miller truck, 1907**

**RENAULT TRUCKS' GENESIS** is connected to two pioneers of the motor industry, Louis Renault and Marius Berliet. The two inventors and entrepreneurs started building cars before turning to truck production, launching their first models in 1900 and 1906, respectively.

During World War I, Renault and Berliet made trucks for the French Army. The wartime demand necessitated major investment in Berliet's factory to increase both space and production capacity. Postwar, the demand for trucks declined greatly, and so both companies scaled back production. Berliet manufactured just one model – the CBA.

**Berliet Stradair**
The Stradair was Berliet's entry into the small truck market. It had striking looks but did not prove popular – only 3,000 were sold before the model was retired in 1970.

Innovation continued at Renault and Berliet during the inter-war years, and both companies introduced models with a diesel engine in 1931. Berliet also produced passenger cars until 1939, but after then focused only on commercial vehicles. While both companies were heavily bombed during World War II, they survived and resumed production after peace was declared. Berliet's first new postwar product was its GLR truck in 1949. In 1950, Renault began producing a small range of trucks and buses, all with a 105-hp engine. Unfortunately, these trucks were not particularly successful.

The postwar era was an active period for mergers and acquisitions in the French truck sector. Between 1952 and 1974, Berliet acquired three manufacturers: Laffly, Rochet Schneider, and Camiva. In 1967, Berliet was bought by Citroën, which itself was owned by Michelin.

**Berliet T100, 1960**
When it was launched, the T100 was the biggest truck in the world, with an engine of 600 hp, but later models had up to 700 hp.

The name "Renault" disappeared for a time from trucks in 1955, when the company's truck division was merged with rivals Latil and Somua to form the Société Anonyme de Véhicules Industriels et d'Équipements Mécaniques – generally known by the acronym "Saviem".

The new group's models were originally badged Saviem-LRS (LRS referring to the founding members), until it became a subsidiary of Renault in 1959, when the initials were dropped. Its first models included a renovated JL range of medium and heavy trucks that had originally been produced by Somua. Saviem launched its first, all-new trucks, the S-range of heavy- and medium-duty trucks, in 1964.

In 1963, Saviem entered a cooperative agreement with German truckmaker MAN, which saw Saviem supply cabs to MAN in return for axles and engines. This partnership produced the SM and JM ranges in France, badged as Saviem-MAN. The collaboration lasted until 1977. Another collaboration was agreed in 1971, when Saviem joined an alliance of European truck manufacturers called the "Club of Four" to develop a shared range of light trucks alongside Volvo, DAF, and Magirus-Deutz.

# "[Louis Renault was] rich, **powerful,** and famous, cantankerous, **brilliant,** often **brutal,** the little Napoleon of an automaking **empire.**"

*TIME,* 1965

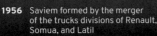

**Berliet CBA**

**Berliet GLR 8**

**Saviem SM 8**

**Renault Trucks E-tech C**

**1900** Louis Renault manufactures his first utility vehicle
**1906** Marius Berliet builds his first truck
**1914** Berliet CBA launches, which becomes very popular with the French Army during World War I
**1931** Both Renault and Berliet introduce diesel engines to their trucks
**1939** Berliet stops production of cars to focus on commercial vehicles
**1949** Berliet launches first postwar product: the GLR truck

**1956** Saviem formed by the merger of the trucks divisions of Renault, Somua, and Latil
**1957** Berliet launches T100, then the largest truck in the world
**1959** Saviem becomes a wholly owned subsidiary of Renault
**1963** Cooperation with German manufacturer MAN on cabs, engines, and axles
**1964** Saviem's first new trucks, the S range of medium trucks, launched
**1965** Saviem buys Richard Continental

**1967** Saviem-MAN SM range launches
**1967** Berliet is bought by Citroën (itself owned by Michelin)
**1971** Saviem becomes one of the "Club of Four" manufacturers collaborating on light trucks
**1974** Renault acquires Berliet from Michelin
**1975** Saviem buys Sinpar
**1978** Berliet and Saviem merge to form Renault Véhicules Industriels
**1983** Renault acquires Dodge Europe
**1990** Acquisition of US Manufacturer Mack

**1990** Launch of Renault Magnum
**1991** Magnum wins International Truck of the Year award
**2001** Renault becomes part of the Volvo Group
**2002** Renault Trucks is formally constituted
**2013** New range of trucks launched – C, D, K, and T
**2020** Launch of first battery-electric model in the Renault Trucks E-tech family

A major success of this project was a shared cab design – often one of the most expensive parts of a truck to design and build – bringing the benefits of economies of scale. From 1975, the trucks that resulted from the alliance were built by Saviem – later Renault – right up until 2001. Some of these cabs made

it to North America, where they were used in the Mack Mid-Liner or Manager.

The Renault name returned to truck grilles in 1978, when Saviem merged with Berliet, which Renault had purchased in 1974. This created Renault Véhicules Industriels – the only truck manufacturer in France.

Renault's bosses then looked to build an international truck group, leading to the acquisition of Dodge Europe in 1983 and iconic US brand Mack in 1990. But its international ambitions also made it a target, and in 2001 Renault became part of the Volvo Group, and Renault Trucks launched as a global brand a year later. One of the most famous models in Renault Trucks' line-up at the time was the Magnum, which was launched in 1990 and was produced until 2013, with upgrades and

**Advertisement for Berliet, 1934**
Berliet and Renault continued to innovate in the 1930s and were both early adopters of the diesel engine.

redesigns along the way. The truck was an instant success – it took the International Truck of the Year title in 1991 and became much loved, especially for its long-haul work. It was replaced in 2013 by the heavy-duty T-series. They are now part of Renault Trucks' family of models that also includes the D, which is aimed at the heavyweight distribution sector, while the C and K series serve the construction industry.

Renault Trucks is currently looking to the future, and has released battery-electric versions of every model in its range, including the 44-tonne (43-ton) Renault Trucks E-Tech T-series. The company is aiming for electric models to make up 50 per cent of its sales by 2030 and to become carbon neutral by 2040.

**Renault Magnum**
The Magnum evolved from Renault's AE series, launched in 1990. They were the first trucks in Europe to feature a flat-floor tractor, making it easier for drivers to move around the cab.

# European Delivery Trucks

The booming inter-war period saw a surge in power, speed, and payload capacity for all types of truck. They also started to embody their own sense of style with new bold and eye-catching liveries. Smaller vehicles developed in France allowed businesses around the globe to get mobile at lower cost. Mass production and cheaper vehicles meant the truck was now shifting the balance in rural regions that had previously relied on horses. Advances in design and manufacturing made it easier to carry out maintenance, while the arrival of gearboxes with synchromesh made driving that much simpler and faster.

Roof made from wooden slats covered in fabric

### △ Ford Model TT

**Date** 1924 **Origin** UK/USA

**Engine** 4-cylinder petrol, 20 hp

**Payload** 1 tonne (1 ton)

The TT enhanced the standard Model T car chassis with an increased load capacity. It proved popular across the world with companies both large and small.

### ▽ Renault MY

**Date** 1924 **Origin** France

**Engine** 4-cylinder petrol, 13.9 hp

**Payload** 1 tonne (1 ton)

Renault's dated coal-scuttle bonnet design soldiered on into the early 1930s. France loved smaller trade vehicles such as this MY, which had the flexibility of both a car and truck, and appealed massively to rural workers.

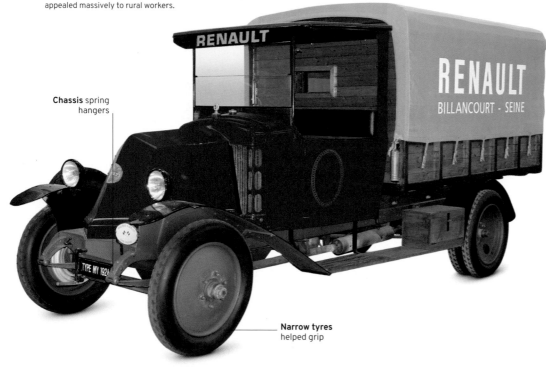

Chassis spring hangers

Narrow tyres helped grip

### △ Austin Heavy 12/4 Delivery Truck

**Date** 1926 **Origin** UK

**Engine** 4-cylinder petrol, 12 hp

**Cylinders** 500 kg (1,102 lb)

The car-based Austin 12 offered exceptional durability, and both van and pickup versions were available from various coach builders. This light brewery delivery truck would have been ideal for negotiating narrow country lanes.

### ▽ Leyland Beaver

**Date** 1933 **Origin** UK

**Engine** 4-cylinder petrol, 74 hp

**Payload** 7.6 tonnes (7.5 tons)

One of the first heavy trucks built for the mainstream market, the Beaver came in both conventional (shown here) and cab-over form. They partly competed with the railways for freight.

### △ Praha Piccolo

**Date** 1932 **Origin** Czechoslovakia

**Engine** 4-cylinder petrol, 16 hp

**Payload** 250 kg (551 lb)

Praha offered the car-based Piccolo as an affordable light pickup. It built a larger LN model from 1932. The Czech firm earned a reputation for quality and took part in expeditions to prove its trucks' endurance.

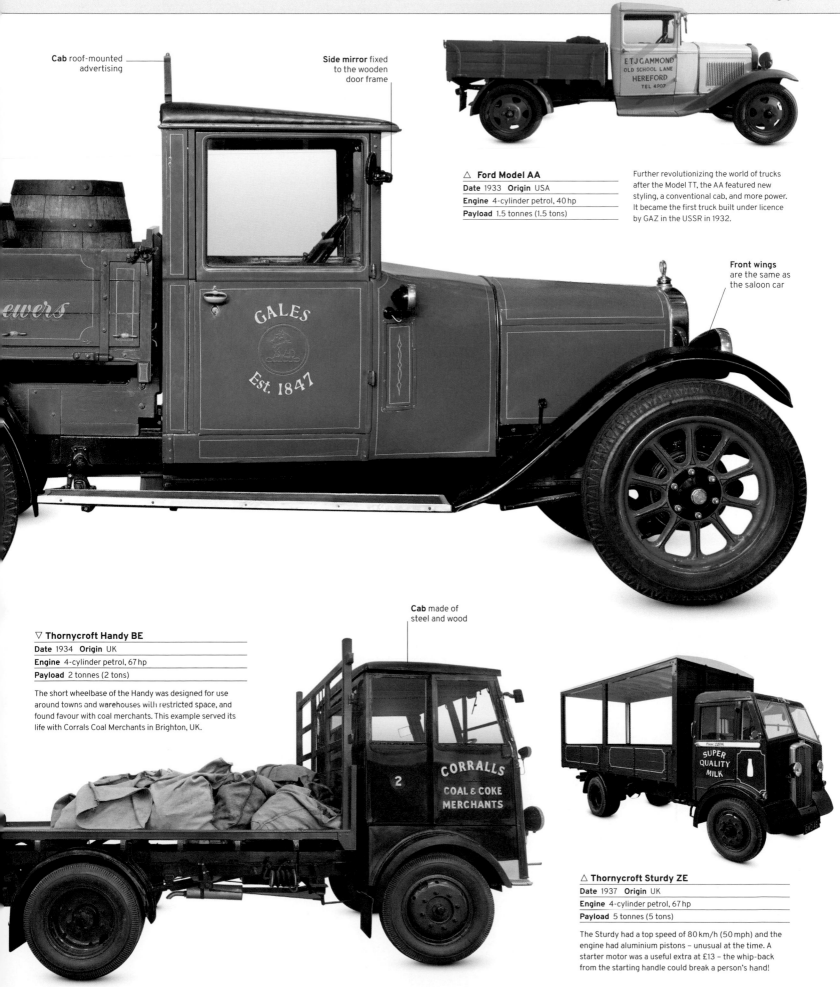

Cab roof-mounted advertising

Side mirror fixed to the wooden door frame

GALES Est. 1847

△ **Ford Model AA**

Date 1933  Origin USA

Engine  4-cylinder petrol, 40 hp

Payload  1.5 tonnes (1.5 tons)

Further revolutionizing the world of trucks after the Model TT, the AA featured new styling, a conventional cab, and more power. It became the first truck built under licence by GAZ in the USSR in 1932.

Front wings are the same as the saloon car

Cab made of steel and wood

▽ **Thornycroft Handy BE**

Date 1934  Origin UK

Engine  4-cylinder petrol, 67 hp

Payload  2 tonnes (2 tons)

The short wheelbase of the Handy was designed for use around towns and warehouses with restricted space, and found favour with coal merchants. This example served its life with Corrals Coal Merchants in Brighton, UK.

CORRALLS COAL & COKE MERCHANTS

SUPER QUALITY MILK

△ **Thornycroft Sturdy ZE**

Date 1937  Origin UK

Engine  4-cylinder petrol, 67 hp

Payload  5 tonnes (5 tons)

The Sturdy had a top speed of 80 km/h (50 mph) and the engine had aluminium pistons – unusual at the time. A starter motor was a useful extra at £13 – the whip-back from the starting handle could break a person's hand!

# Ford Model TT

The first Ford Model T car rolled off the Michigan production line in 1908, and by the mid-1920s half of the world's registered motorcars were Fords. Introduced in 1917, the Model TT truck featured a longer and tougher chassis frame than its car predecessor. It also had a stronger rear axle, while various gearbox options were available from independent suppliers that could enhance the ability of this rugged, dependable truck.

**THE MODEL TT** was Ford's first purpose-built truck, rather than an adapted car-based vehicle, but even so it shared many features with the Model T. The TT was priced from $600 at its launch, which was almost twice the cost of the car version, but both the engine and the cabin were the same, with the fuel tank still situated under the seat. New features included a longer, stronger chassis, which could support a payload of around a tonne. That capacity proved useful to many industries, and TT variants ranged from flatbeds and panel trucks, to buses and tankers. The truck was initially only available as a chassis and engine until 1924, when Ford began making its own bodies. Ford advised a maximum speed of 32 km/h (20 mph), but many owners talk of achieving higher speeds from the 20 hp engine.

**Metal footplate**
Branded cast-alloy footplates assist with grip when getting in and out of the cab. These were necessary because the Model TT's cab is higher than that of the Model T car.

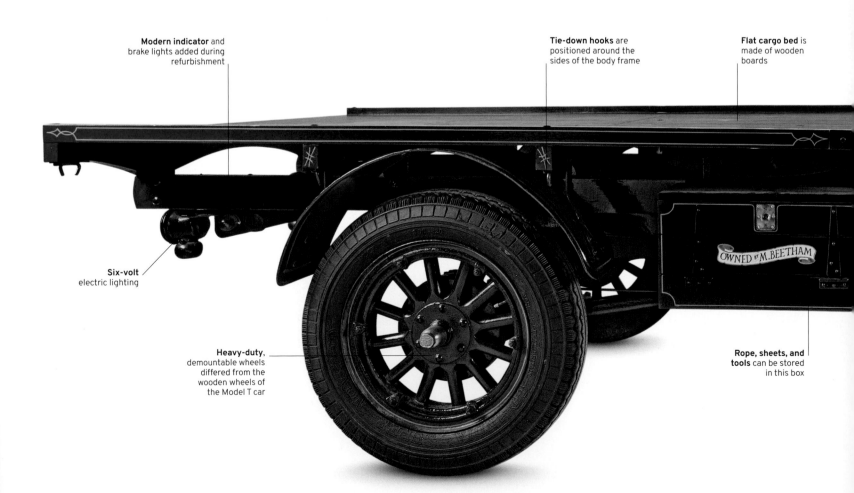

**Modern indicator** and brake lights added during refurbishment

**Tie-down hooks** are positioned around the sides of the body frame

**Flat cargo bed** is made of wooden boards

**Six-volt** electric lighting

**Heavy-duty,** demountable wheels differed from the wooden wheels of the Model T car

OWNED BY M. BEETHAM

**Rope, sheets, and tools** can be stored in this box

## SPECIFICATIONS

| | |
|---|---|
| Model | Ford Model TT |
| Origin | USA |
| Built | 1924 |
| Assembly | Detroit, US and Manchester, UK |
| Production run | 1917–28, approx. 1.3 million |
| Weight | Dependent on body |
| Load capacity | 0.9 tonnes (0.9 tons) |
| Engine | 4-cylinder, 2898 cc, 20 hp, |
| Transmission | Two-speed planetary – other gearbox options were available |
| Maximum speed | 64 km/h (40 mph) |

**FRONT VIEW**

**REAR VIEW**

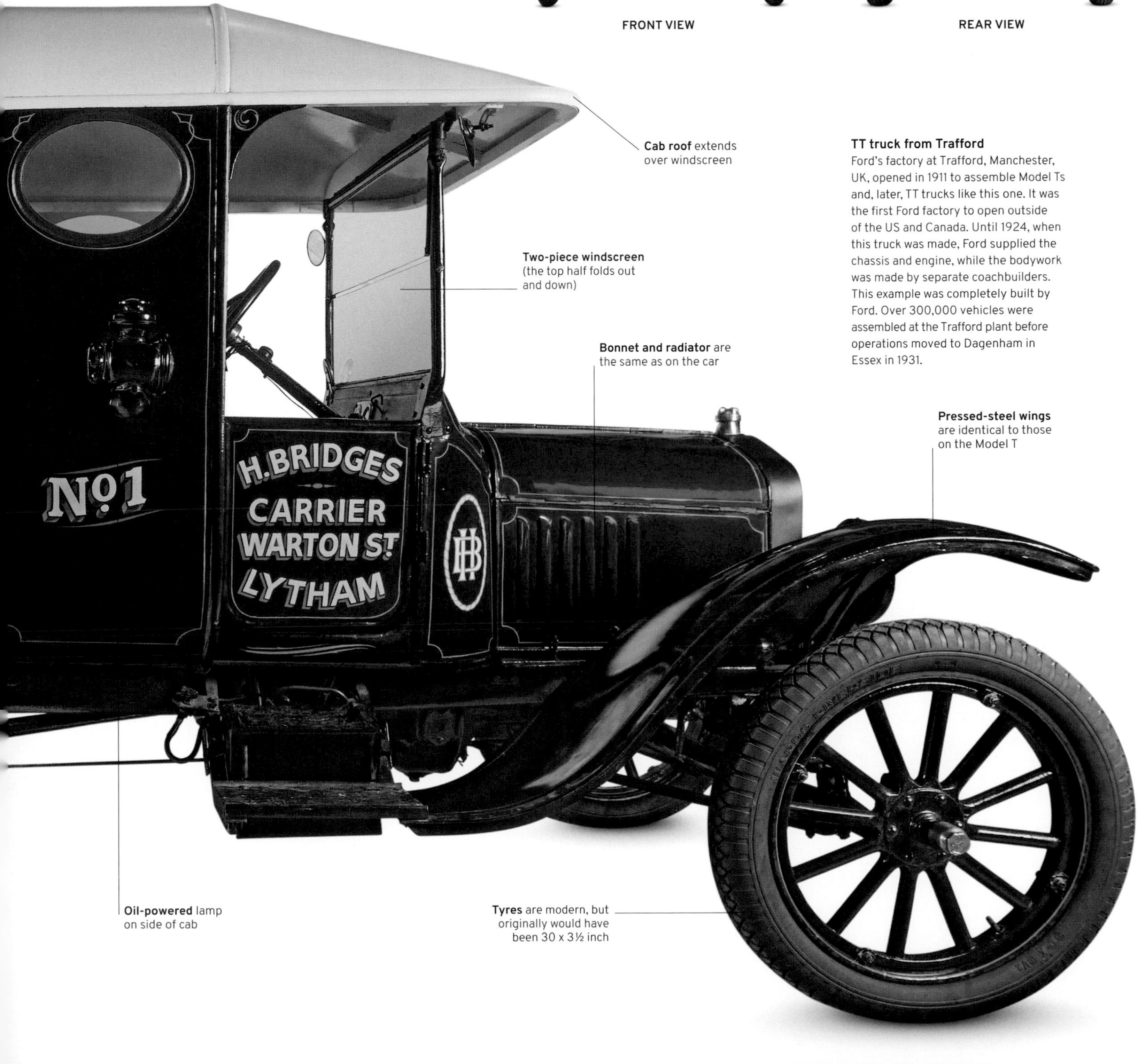

**Cab roof** extends over windscreen

**Two-piece windscreen** (the top half folds out and down)

**Bonnet and radiator** are the same as on the car

**Oil-powered** lamp on side of cab

**Tyres** are modern, but originally would have been 30 x 3½ inch

**TT truck from Trafford**

Ford's factory at Trafford, Manchester, UK, opened in 1911 to assemble Model Ts and, later, TT trucks like this one. It was the first Ford factory to open outside of the US and Canada. Until 1924, when this truck was made, Ford supplied the chassis and engine, while the bodywork was made by separate coachbuilders. This example was completely built by Ford. Over 300,000 vehicles were assembled at the Trafford plant before operations moved to Dagenham in Essex in 1931.

**Pressed-steel wings** are identical to those on the Model T

H. BRIDGES
CARRIER
WARTON ST.
LYTHAM

No.1

## THE EXTERIOR

Throughout the production life of the Model TT truck, many different body styles were used. This particular example features a traditional English-style flatbed body that was popular with coal merchants and farmers. Amazingly, it survived long enough to be restored, and in 1995 the owners brought this Model TT back into show-winning condition.

**1.** The Model TT, like the car, was often referred to as "Tin Lizzie"; the origins of the nickname are uncertain, but it could have been taken from a famous race car of the period known as "Old Liz" **2.** Livery motif **3.** Brass radiator cap, used to top up with water when required, are commonplace for the era **4.** Carriage-style, oil-powered side lamps were fitted either side of the scuttle **5.** Detachable wheel rims for changing tyres **6.** This particular body has a built-in tool box that would have been used to store rope for securing loads

## THE INTERIOR

Internally, the cab is striking in its simplicity. The two small levers attached to the steering wheel are the throttle control and the spark advance, which controls spark timing. There are three floor pedals: on the left is a clutch pedal that engages the gears, the middle pedal is a reverse pedal, and the far right pedal acts as the brake. A parking brake/gear-release floor lever is to the right of the steering wheel. The ammeter gauge, that measures current flow, and ignition switch are mounted on a small panel. TTs were designed to be even more spartan than the cars: with some coachbuilders, the seat was little more than a piece of wood over the fuel tank.

**7.** Basic steering wheel controls; a Model TT has surprisingly sensitive steering **8.** Ignition switch and ammeter charging gauge are all that is fitted. Optional aftermarket speedometers were available **9.** Pedal layout is the same as the Model T

## THE ENGINE

Unique when its was introduced in 1908, the engine of the Model T was reused in the TT truck. All four cylinders were cast as a single block. Also incorporated into this cast-iron "monobloc" is the crankcase, avoiding the expense of having a separate aluminium casing. This enabled engines to be assembled more quickly, thereby cutting costs and reducing the price to the customer. The engine could typically endure decades of relentless, rigorous work.

**10.** Engine, showing the 6-volt electric horn (the horizontal wired cylinder)
**11.** The driver's side of the engine showing the Kingston carburettor (bottom left corner), as fitted to every Model T car and TT truck

# First Semi-trailers

To meet the increase in demand for goods transported by road after World War I, a new type of trailer was developed to speed up delivery and carry greater loads, using existing trucks fitted with a coupling device, known as the "fifth wheel". This semi-trailer – or semi-articulated trailer – was invented by the Fruehauf Trailer Company of Detroit in 1914. The trailer had no front axle, and could be quickly coupled and decoupled from the truck via the fifth wheel. Such trailers offered greater flexibility, bearing two to three times the weight that the chassis of a truck alone could support. Trailers could also be much longer than flatbed wagons, as their articulation made them more manoeuvrable.

**Short front** for easy manoeuvring

### △ Mack AC "Bulldog"

| | |
|---|---|
| **Date** 1925 | **Origin** USA |
| **Engine** 4-cylinder petrol, 50 hp | |
| **Payload** Variable | |

Introduced in 1916, the AC earned the "Bulldog" name for its tenacious work in World War I. Its solid tyres were an advantage in places where punctures could be frequent, such as building sites.

### ▽ International A-5

| | |
|---|---|
| **Date** 1930 | **Origin** USA |
| **Engine** 6-cylinder petrol, 65 hp | |
| **Payload** 3 tonnes (3 tons) | |

One of the earliest and most successful tractor trucks, the wood-framed, fabric-topped A-5 towed a variety of trailers, and was a stalwart of heavy-duty farm work. The powerful engine was driven by a six-speed gearbox, which was unusual in this size of truck.

SCAMMELL LORRIES LTD.,
TOLPITS LANE,
WATFORD, HERTS.
TEL.5231 (5 LINES)

### △ Scammell Mechanical Horse

| | |
|---|---|
| **Date** 1934 | **Origin** UK |
| **Engine** 4-cylinder petrol, 38.5 hp | |
| **Payload** 3 tonnes (3 tons) | |

Scammell pioneered the small articulated vehicle with its three-wheeled Mechanical Horse. The truck's ability to make tight 360-degree turns made it popular with breweries and railway companies transporting loads in confined locations.

**Double strip** chrome bumper

**Fifth-wheel** coupling with electric contacts for trailer lighting

### ▷ Diamond T 412DR

| | |
|---|---|
| **Date** 1936 | **Origin** USA |
| **Engine** 6-cylinder petrol, 118 hp | |
| **Payload** 18 tonnes (17.7 tons) | |

The 412DR was a move into heavy trucks for Diamond T. The tractor unit pictured pulled a flatbed trailer with a box body. The tall exhaust pipe gives the impression of a much larger rig.

### △ Ford BB

**Date** 1937  **Origin** USA

**Engine** V8 petrol, 85 hp

**Payload** 1.4 tonnes (1.4 tons)

The BB truck was part of a light truck generation that used Ford's powerful flathead V8 engine. With its double-wheel rear axle, the BB was used to pull smaller semi-trailers, such as tankers. Its chain-driven engine offered greater power than belt-driven alternatives.

**Streamlined cab** based on the Studebaker passenger cars

### △ Studebaker J30M 4×2

**Date** 1937  **Origin** USA

**Engine** 6-cylinder petrol, 98 hp

**Payload** 2.7 tonnes (2.7 tons)

The largest of the Studebaker J-series trucks, the J30M featured a heavy-duty front axle and effective brakes that helped to make it an ideal semi-trailer tractor truck. Studebaker trucks, with their distinctive curved cab, would go on to serve the US Army during World War II.

**Coupling** is located under cargo box

**Quick-release** trailer coupling gear

### ▽ Mack BX

**Date** 1939  **Origin** USA

**Engine** 6-cylinder petrol, 128 hp

**Payload** 5.4 tonnes (5.4 tons)

The BX, the largest pre-World War II Mack truck, was used throughout the US as a tractor unit in large industrial and commercial delivery fleets. Wind-down legs in the middle of the trailer were used when decoupling from the truck.

### ▽ Mack EHU

**Date** 1939  **Origin** USA

**Engine** 6-cylinder petrol, 100 hp

**Payload** 4.5 tonnes (4.5 tons)

In the 1930s, tractor unit trucks like the EHU, with a flat-fronted cab sitting above the front axle – known as "cab-overs" – were relatively rare. Some EHU cabs featured a sleeper cabin for long-distance runs. The truck saw service with both the US and British armies in World War II.

**Fifth-wheel** trailer coupling

MAN production line
in Nuremberg, 1929

# Key Manufacturers
# The MAN Story

German manufacturer MAN has been producing trucks for more than 100 years. The company's first light and heavy commercial vehicles came out in 1915, and it soon had ranges of tractors, buses, and trucks. Respected across Europe, MAN has gained global recognition with multiple International Truck of the Year awards.

**MAN CAN TRACE** its roots back more than 250 years, but the company as we know it today came about though the merger of sister engineering businesses Maschinenfabrik Augsburg and Maschinenbau-Aktiengesellschaft Nürnberg, in 1898. Renamed Maschinenfabrik Augsburg-Nürnberg AG, or MAN for short, in 1908, the company produced its first truck in 1915. Initially, production was at the MAN-Saurer Truck Works in Lindau on Lake Constance – a joint venture with Swiss business Adolph Saurer AG, before production moved to Nuremberg in 1916. In 1928, MAN developed a 150 hp six-wheeled truck. With a petrol engine and a capacity of 10 tonnes (9.8 tons), this became a prototype for the company's future heavy-duty models. It was followed in 1932 by the S1H6, powered by MAN's D4086 engine that produced 140 hp, and was the world's most powerful diesel truck at the time. Another 10 hp was added a year later as diesels came to the fore in their range.

As World War II approached, MAN switched focus to developing military vehicles, although it continued to innovate, with the introduction of all-wheel drive. To meet the postwar demand for trucks, MAN launched its

**MAN diesel truck poster, 1949**
MAN launched its MK range with 110–120 hp engines shortly after WWII, when there was a huge demand for trucks necessary for postwar reconstruction.

MK range, powered by 6-cylinder 110–120 hp diesel engines.

In 1951, MAN became the first German truck manufacturer to introduce turbocharging, which utilized exhaust gases to boost engine power. This provided a 35 per cent increase in performance over its rivals' trucks with naturally aspirated engines, boosting the output of their 6-cylinder engine from 130 to 175 hp. Also popular around this time was MAN's F8 short-nosed truck. With a 180-hp V8 engine and a capacity of 19–22 tonnes (18.7–21.7 tons), it was the most powerful truck in Europe when it was introduced in 1953.

**The first 515 L1, 1955**
This was the first truck to roll off MAN's new larger plant in Munich-Allach in 1955. It had a 115-hp engine and a payload of 5 tonnes (5 tons).

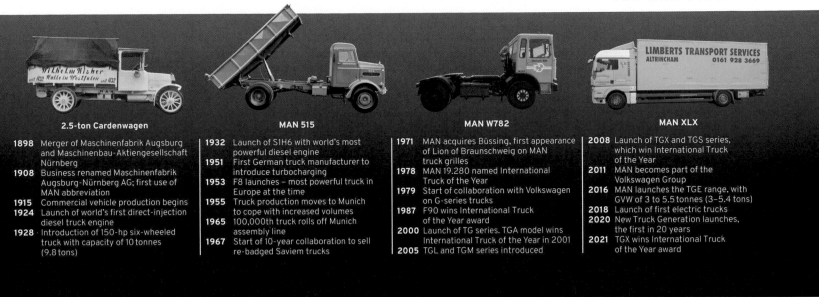

**2.5-ton Cardenwagen**

| | |
|---|---|
| **1898** | Merger of Maschinenfabrik Augsburg and Maschinenbau-Aktiengesellschaft Nürnberg |
| **1908** | Business renamed Maschinenfabrik Augsburg-Nürnberg AG; first use of MAN abbreviation |
| **1915** | Commercial vehicle production begins |
| **1924** | Launch of world's first direct-injection diesel truck engine |
| **1928** | Introduction of 150-hp six-wheeled truck with capacity of 10 tonnes (9.8 tons) |

**MAN 515**

| | |
|---|---|
| **1932** | Launch of S1H6 with world's most powerful diesel engine |
| **1951** | First German truck manufacturer to introduce turbocharging |
| **1953** | F8 launches – most powerful truck in Europe at the time |
| **1955** | Truck production moves to Munich to cope with increased volumes |
| **1965** | 100,000th truck rolls off Munich assembly line |
| **1967** | Start of 10-year collaboration to sell re-badged Saviem trucks |

**MAN W782**

| | |
|---|---|
| **1971** | MAN acquires Büssing, first appearance of Lion of Braunschweig on MAN truck grilles |
| **1978** | MAN 19.280 named International Truck of the Year |
| **1979** | Start of collaboration with Volkswagen on G-series trucks |
| **1987** | F90 wins International Truck of the Year award |
| **2000** | Launch of TG series. TGA model wins International Truck of the Year in 2001 |
| **2005** | TGL and TGM series introduced |

**MAN XLX**

| | |
|---|---|
| **2008** | Launch of TGX and TGS series, which win International Truck of the Year |
| **2011** | MAN becomes part of the Volkswagen Group |
| **2016** | MAN launches the TGE range, with GVW of 3 to 5.5 tonnes (3–5.4 tons) |
| **2018** | Launch of first electric trucks |
| **2020** | New Truck Generation launches, the first in 20 years |
| **2021** | TGX wins International Truck of the Year award |

# On several occasions, MAN has produced the **most powerful** truck in Europe, from **150 hp** in 1928 to **680 hp** in 2008.

With production volume increasing, MAN moved its truck-building operation from Nuremberg to a new and bigger factory in Munich-Allach. The first truck, a 515 L1, rolled off the assembly line on November 1955. Just 10 years later, the 100,000th model was produced there – indicating just how popular the company's trucks were at that time.

MAN developed its international business in 1967, collaborating with French manufacturer Saviem to sell the latter's light- and medium-duty trucks in Germany and other markets, although badged as MAN. This partnership lasted for a decade.

There was more domestic expansion in 1971 when MAN acquired bus and truck manufacturer Büssing, which had been struggling financially. MAN continued to produce some of Büssing's models – with a MAN cab added – badged as MAN-Büssing. This deal also saw MAN adopt Büssing's logo,

the iconic Lion of Braunschweig, which has appeared on the grilles of MAN vehicles ever since.

MAN's international reputation grew throughout the 1970s, culminating in it winning the International Truck of the Year title for the first time in 1978 for its 19.280 model. The following year, MAN collaborated with fellow German marque Volkswagen, using the latter's LT body for its own G-series medium trucks. This deal lasted until 1993. However, MAN still produced its own trucks, and in the mid-1980s launched its G90, M90, and F90 ranges of light-, medium-, and heavy-duty trucks respectively – with the latter winning the International Truck of the Year title in 1987. This line-up of trucks is still used today, albeit in a modernized form.

In 2000, MAN launched its TG series, with its heavy-duty TGA truck, scooping International Truck of the Year award the following year. Its siblings in the light

**MAN D3876 LE12x 6-cylinder engine**

and medium categories, the TGL and TGM models, followed in 2005. These were introduced along with a new generation of more fuel-efficient engines. Production of the TGA ceased in 2008, when it was replaced by the TGX and TGS – which again took the honours at the International Truck of the Year Awards. The trucks could be specified with a V8 680-hp engine, which at the time was the most powerful in Europe.

In 2016, MAN expanded its light commercial offerings with the launch of the TGE range. Two years later, MAN showcased its battery-electric version of the TGE – the eTGE, along with an eTGM, at the prestigious IAA conference for leaders in the logistics and transportation industry. It was

**MAN TGX, 2018**
The TGX is MAN's current tractor unit. First launched in 2007, the truck was given a facelift in 2020, which featured 20,000 new parts.

one of the first manufacturers to launch electric versions of its entire range of trucks with capacities from 3–26 tonnes (3–25.6 tons).

MAN went back to the drawing board for its most recent set of truck launches. The new MAN Truck Generation was launched in February 2020 and was the culmination of 12 million work-hours. It set new industry standards for intelligent driver-assistance systems that improve safety by guiding manoeuvres, such as turns and lane changes.

# Taking Trucks to Market

As the automotive industry took off, vehicle manufacturers realized it was vital to get their latest products in front of potential buyers. One of the first motor shows, the "Horseless Carriage Exhibition", was held in South Kensington, London, in 1896. A decade later saw the debut of the Commercial Motor Transport Show – an event for truck, vans, and buses – at Olympia Exhibition Centre, London.

## TRUCK SHOWS IN THE 21ST CENTURY

These days, rebranded as The Commercial Vehicle Show, the annual event remains the largest road transport sector gathering in the UK, with hundreds of exhibitors showcasing trucks, vans, and related products and services. Europe's largest commercial truck show – the IAA (Internationale Automobil-Ausstellung) Transportation event – made its debut in Germany in 1897 with an exhibition of eight vehicles and achieved a record attendance of 825,000 by 1939. Among the many trucking events in the US, Indiana's Work Truck Show, established in 2001, is often considered the largest. China held its inaugural Commercial Vehicles Show – a biannual event and the biggest exhibition in Asia – in 2012 at the Wuhan International Expo Centre.

**Exhibits from** Renault, Dennis, Daimler, and others, at the 1935 Commercial Motor Transport Show, held at the Olympia Exhibition Hall, London.

# Delivery Vans

During the 1920s and 30s, as business was booming in the US and Europe during the inter-war period, the range of commercial vehicles available was extremely diverse. At their most basic, small pedal-powered "velocars" with van bodies could be bought cheaply in France, while in America, emerging truck firms were producing vans and pickups with a level of refinement that had only previously been available in cars. All of these vehicles shared one thing in common: none of them were hard to drive. Owners chose them to fit with their lifestyle as well as their business.

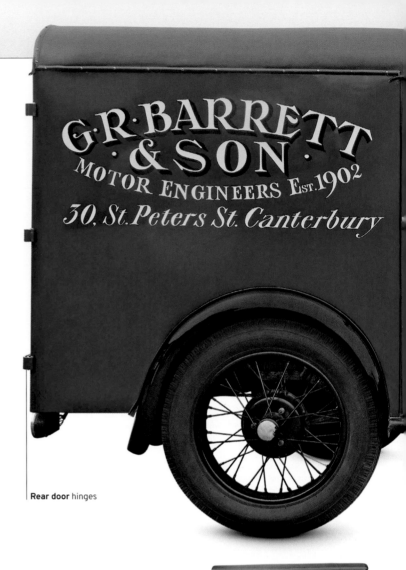

G·R·BARRETT & SON · MOTOR ENGINEERS Est.1902
30, St. Peters St. Canterbury

**Rear door** hinges

Photo courtesy of Hyman Ltd. www.hymanltd.com

**Canvas-covered** roof

**Wooden-framed** rear cargo body

**Cab door**

### △ Ford Model A Postal Van

| | | |
|---|---|---|
| **Date** 1927 | **Origin** USA | |
| **Engine** 4-cylinder petrol, 40 hp | | |
| **Payload** 500 kg (1,100 lb) | | |

This small mail van would have been used to deliver to rural communities where other methods would have taken too long. A single seat and wooden body kept both weight and cost down.

### ▽ Twin Coach

| | |
|---|---|
| **Date** 1931 | **Origin** USA |
| **Engine** 4-cylinder petrol Hercules 37.7 hp | |
| **Payload** 500 kg (1,100 lb) | |

This attractive, hard-working van had a saddle rather than a seat, so that the driver could operate the vehicle standing up in between delivery stops. Twin Coach made a variety of public service vehicles up until 1960.

Helms Bakeries

Helms Bakeries    Daily at Your Door

UPhone O-2141

H    58

Photo courtesy of Hyman Ltd. www.hymanltd.com

ХЛЕБ

### △ GAZ AA Bread Van

| | |
|---|---|
| **Date** 1932 | **Origin** Russia |
| **Engine** 4-cylinder petrol, 50 bhp | |
| **Payload** 1.5 tonnes (1.5 tons) | |

The Soviet truck firm GAZ built their first vehicle under licence from Ford in 1932. The production of these trucks helped spur large-scale manufacturing of a new generation of trucks in Russia.

### △ Fiat 508 Balilla

| | |
|---|---|
| **Date** 1934 | **Origin** Italy |
| **Engine** 4-cylinder petrol, 24 hp | |
| **Payload** 350 kg (770 lb) | |

This van was based on the chassis of the extremely successful 508 car model. The commercial version was made as a flat-bed pickup, which was the basis of many different forms, including vans.

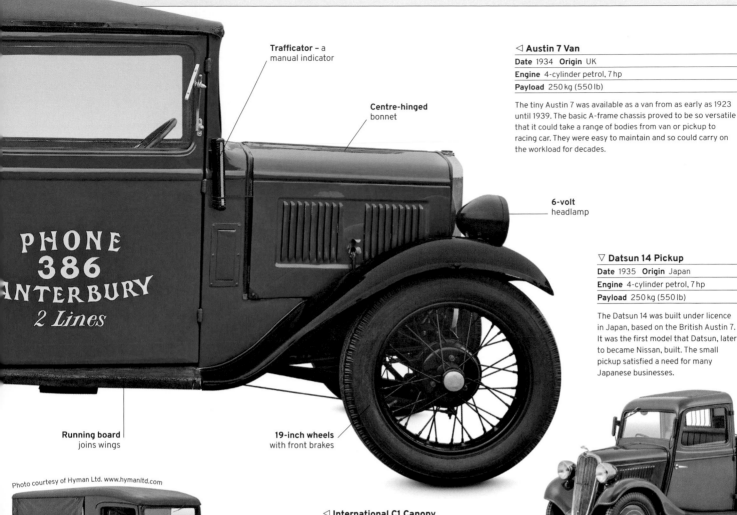

Trafficator – a
manual indicator

Centre-hinged
bonnet

6-volt
headlamp

PHONE
386
ANTERBURY
2 Lines

Running board
joins wings

19-inch wheels
with front brakes

Side mount
spare wheel

### ◁ Austin 7 Van

| | |
|---|---|
| **Date** 1934 | **Origin** UK |
| **Engine** 4-cylinder petrol, 7 hp | |
| **Payload** 250 kg (550 lb) | |

The tiny Austin 7 was available as a van from as early as 1923
until 1939. The basic A-frame chassis proved to be so versatile
that it could take a range of bodies from van or pickup to
racing car. They were easy to maintain and so could carry on
the workload for decades.

### ▽ Datsun 14 Pickup

| | |
|---|---|
| **Date** 1935 | **Origin** Japan |
| **Engine** 4-cylinder petrol, 7 hp | |
| **Payload** 250 kg (550 lb) | |

The Datsun 14 was built under licence
in Japan, based on the British Austin 7.
It was the first model that Datsun, later
to became Nissan, built. The small
pickup satisfied a need for many
Japanese businesses.

Box
loading
area

### ◁ International C1 Canopy Express

| | |
|---|---|
| **Date** 1936 | **Origin** USA |
| **Engine** 4-cylinder petrol, 78 hp | |
| **Payload** 0.9 tonne (0.9 ton) | |

One of the smaller models produced by
this well-known heavy truck manufacturer,
the C1 Canopy Express was light and agile,
and suited a wide range of trades.

### ▽ Nissan Type 80 Van

| | |
|---|---|
| **Date** 1937 | **Origin** Japan |
| **Engine** 6-cylinder petrol, 84 hp | |
| **Payload** 2 tonnes (2 tons) | |

Loosely based on the Graham-Paige
saloon car, the Type 80 was Nissan's
first cab-over truck. This example has
undergone full restoration back to its
Miksukoshi department store livery.

Extra windows
above windscreen
offer more light

### △ Renault AG C2

| | |
|---|---|
| **Date** 1938 | **Origin** France |
| **Engine** 4-cylinder petrol, 85 hp | |
| **Payload** 2 tonnes (2 tons) | |

Up until the late 1930s, Renault's main commercial
vehicle output was under 1 tonne. The larger AG C2
featured twin rear wheels; 6,256 were produced.

# Leyland Beaver

Leyland had won a reputation for dependability through supplying trucks during World War I. Departing from its sombre battle-ready image, Leyland brought out a new range of trucks from the late 1920s, each model bearing a light-hearted animal name, including the Lioness, Tiger, and Beaver. Introduced in 1928, the Beaver became a popular flatbed truck and found use in a wide range of industries.

| SPECIFICATIONS | |
|---|---|
| Model | Leyland Beaver |
| Origin | UK |
| Assembly | Lancashire |
| Production run | Unknown |
| Weight | 5 tonnes (5 tons) |
| Payload | 3 tonnes (3 tons) |
| Engine | 4-cylinder petrol (diesel 5.7 fitted) |
| Transmission | 3-speed crash gearbox |
| Maximum speed | 56 km/h (35 mph) |

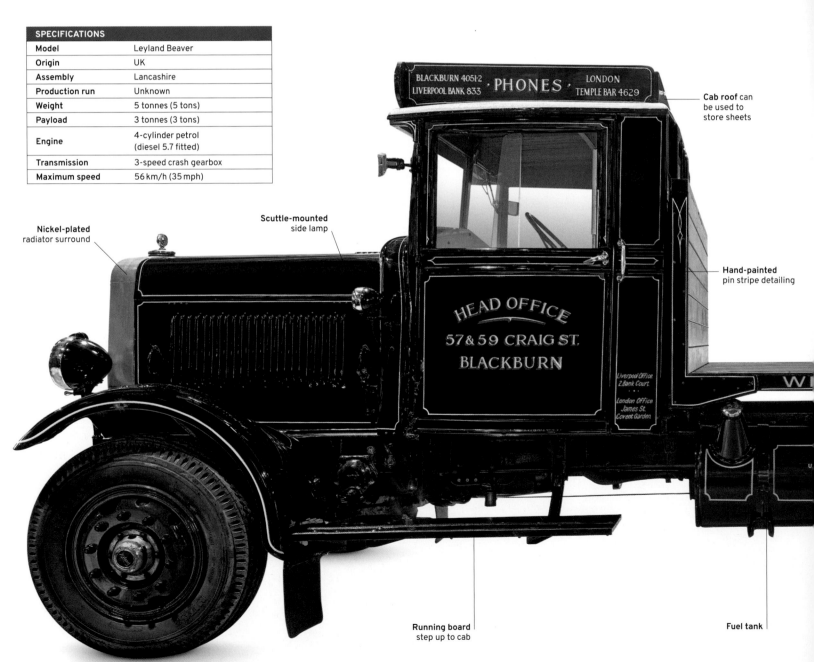

**Cab roof** can be used to store sheets

**Hand-painted** pin stripe detailing

**Nickel-plated** radiator surround

**Scuttle-mounted** side lamp

**Running board** step up to cab

**Fuel tank**

**THE LEYLAND BEAVER'S CHASSIS**, like those of other manufacturers, could be used by coachbuilders to underpin a wide range of bespoke vehicles, such as vans, buses, and fire engines. However, the flatbed model seen here is exactly as it would have been when it left the Leyland Factory. On the earliest Beaver models, as here, the engine sits ahead of a rudimentary cab that offered little comfort and was also prone to leaking. By the mid-1930s, a "cab over" version was available, with the cab sitting above the engine. This design created a longer load area and eventually became prevalent throughout Europe. Other improvements included larger tyres for increased comfort and load capacity. The large flat radiator, however, remained unchanged across the range. The Leyland Beaver came in many other forms, including tankers, drop-side trucks, and even as tractor units for articulated trailers.

**Leyland badge**
Radiator badges were the easiest way of advertising a vehicle, be it a car or truck. This Beaver radiator badge is typical of the period.

**FRONT VIEW**

**REAR VIEW**

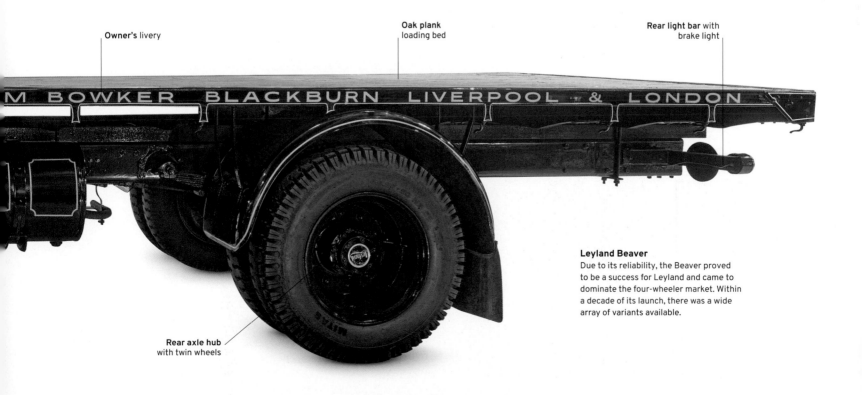

**Owner's** livery

**Oak plank** loading bed

**Rear light bar** with brake light

**Rear axle hub** with twin wheels

**Leyland Beaver**
Due to its reliability, the Beaver proved to be a success for Leyland and came to dominate the four-wheeler market. Within a decade of its launch, there was a wide array of variants available.

## THE EXTERIOR

Two versions of the Beaver were offered: a bonnet model, like the one pictured here, and a cab-over version that allowed for a slightly longer loading bed. At just over 1.2 m (4 ft) from the ground to the cab floor, the Beaver was a tall truck by early 1930s standards. In 1933, this vehicle was delivered new to Crown Wall Coverings Ltd. of Darwen, Lancashire, and remained in service until 1960. Decades later, it was discovered in a field in Dorset and the new owner carried out a full restoration. The current livery represents a haulage firm that used to run similar trucks to this example.

**1.** Hand-painted livery of the owner/operator **2.** Cast-alloy radiator badge **3.** Headlight **4.** Radiator-mounted temperature gauge, which can be viewed from within the cab **5.** Cab-mounted side lamp **6.** Steering arm connecting steering box to front axle **7.** Fuel tank

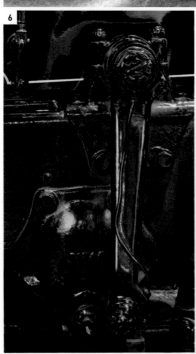

## THE INTERIOR

A mahogany dashboard greets the driver upon entering the small cab. In the centre of the cab is the gear change: up until this point, it was conventional with most trucks to have the gear change placed awkwardly on the right-hand side of the floor. Despite being such a large truck, the interior space is quite cramped, and virtually no thought has been given to the driver's comfort.

**8.** Interior of the cab  **9.** Chassis plate displaying hand-stamped identification numbers  **10.** Dashboard panel with speedometer, clock, oil pressure gauge, and ammeter  **11.** Fuel pump for engine

## THE ENGINE

Originally, this Beaver truck would have been equipped with a four-cylinder petrol engine with cast-alloy engine parts, which had become commonplace by the 1920s. The engine seen here, however, is an aftermarket replacement fitted in 1940, when petrol rationing was introduced. This 6-cylinder, 5.7-litre Leyland diesel engine was also used in Leyland buses and power boats of the period.

**12.** Steering box (bottom)  **13.** Leyland 5.7-litre diesel engine

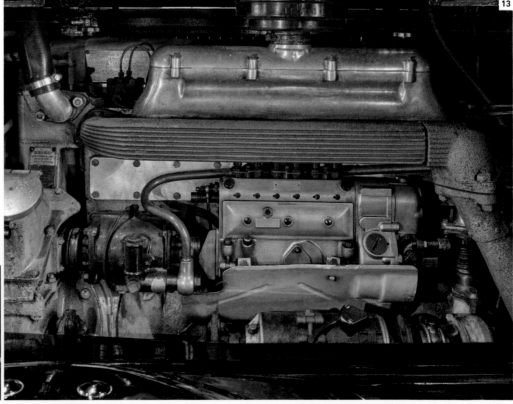

# Trucks in Transition

Whilst Art Deco and big bands picked up the pace in the 1930s, heavy trucking evolved in the shadows without the need for glamour or praise. As trucks became cheaper and more reliable they were used more widely in a range of industries, leading to new models being introduced. Trucks of the previous decades were little more than leftovers from World War I – slow, heavy, and tired. Now trucks had pneumatic tyres, enclosed cabs, sprung seats, heaters, and for some, the biggest advancement: diesel engines. While truck design was advancing quickly, older technology persisted. The ready availabilty of coal and water meant that steam was still a viable choice.

### △ Foden 5-ton HH Wagon

| | |
|---|---|
| **Date** 1930 | **Origin** UK |
| **Engine** Steam compound | |
| **Payload** 5.1 tonnes (5 tons) | |

This steam wagon had its engine mounted above the boiler, making for a cramped cab area. Future models moved the engine under the chassis to create more room. Like most steam wagons in the 1930s, it was fitted with pneumatic tyres.

**Same engine**
used throughout
the range

### ◁ Thornycroft A2 Cattle Wagon

| | |
|---|---|
| **Date** 1930 | **Origin** UK |
| **Engine** 6-cylinder petrol, 33 hp | |
| **Payload** 2 tonnes (2 tons) | |

The A2 was very popular for transporting livestock, and many were used as horse boxes. Also, because Thornycroft had proven its reliability serving the British army, the UK government offered a subsidy to buyers. Cab-over types were also available.

### △ Foden 4½ NHP C Type

| | |
|---|---|
| **Date** 1931 | **Origin** UK |
| **Engine** Overtype steam, 4 hp | |
| **Payload** 6.1 tonnes (6 tons) | |

By 1934, Foden had discontinued production of its steam wagons in favour of turning their focus to internal combustion driven trucks. Despite this, the C type retained a loyal following amongst showman drivers. Many were updated with pneumatic tyres for long distance travelling, while others were cut down into short timber tractors.

**Windscreen**
opens

### ▷ Fiat 634N

| | |
|---|---|
| **Date** 1933 | **Origin** Italy |
| **Engine** 6-cylinder diesel, 75 hp | |
| **Payload** 7.6 tonnes (7.5 tons) | |

Manufactured up until 1939, the Fiat 634 was the largest truck manufactured in Italy up to this point, popular both at home and abroad. Uprated military versions proved to be vital logistic support throughout WWII and beyond.

Chrome-plated
radiator shell

◁ **Mercedes-Benz Lo 2750**

| | |
|---|---|
| **Date** 1933 | **Origin** Germany |
| **Engine** 4-cylinder diesel, 65 hp | |
| **Payload** 2.75 tonnes (2.7 tons) | |

The first truck to be fitted with a diesel engine as standard, this long-running and long-bonnetted series of Mercedes trucks were among some of the finest quality vehicles available in the world. This example was used to transport the legendary Silver Arrows Grand Prix racing cars of the 30s to race meetings.

Glamorous
radiator and
cab styling

▷ **Mack CJ Dump Truck**

| | |
|---|---|
| **Date** 1933 | **Origin** USA |
| **Engine** 6-cylinder petrol, 128 hp | |
| **Payload** 5.4 tonnes (5.3 tons) | |

Mack built 79 of these CJ model trucks from 1933 until 1941. The company had a reputation within the construction industry for robust trucks and heavy dumpers, like this sturdy workhorse.

Twin-wheeled rear
axles for greater
load capacity

△ **Isotta Fraschini D80**

| | |
|---|---|
| **Date** 1939 | **Origin** Italy |
| **Engine** 6-cylinder diesel, 95 hp | |
| **Payload** 6.5 tonnes (6.4 tons) | |

Although best known for their motorcars fit for heads of state, Isotta entered into truck manufacturing with the D80 in 1934. Perhaps one of the most attractive trucks of the period, with its distinctive curved grille, its design was years ahead of its competition.

**TALKING POINT**

## Mercedes-Benz O 10000

The Austrian postal service commissioned 386 mobile post buses in 1938. Mercedes-Benz drew upon all its experience in truck manufacturing to create the unusual vehicle. The bus had six wheels and was 14 m (46 ft) long, functioning as a community hub wherever it stopped. In addition to sending and collecting letters and parcels from it, customers could make calls from one of the bus's three phone kiosks. Such was the quality of these vehicles that this one remained in service for over 40 years.

**The O 10000** was the largest Mercedes bus of the 1930s. It was powered by a six-cylinder diesel engine.

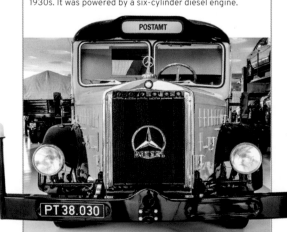

# Lights, Camera, Traction!

In the early days of Hollywood, ingenuity, creativity, and enthusiasm made up for a lack of health and safety measures. Like many industries in the 1920s, US film makers were quick to take advantage of the versatility of trucks – in this case, even as mobile sets.

### DOING IT FOR REAL
Dorothy Devore's film career began in 1918 and ran until 1930, just as movies with soundtracks were starting to gather pace. She was primarily known for madcap stunt-based comedies that often saw her risking life and limb (in those days for real) all for the entertainment of her audience. Rear projection techniques were still several years away, so if a scene required movement, then everybody needed to actually move. In an era of Hollywood that saw asbestos commonly used for fake snow, this was most probably a risky undertaking.

**American actor Dorothy Devore** lies in a bed that is strapped to a flatbed truck while a camera crew films a dream sequence.

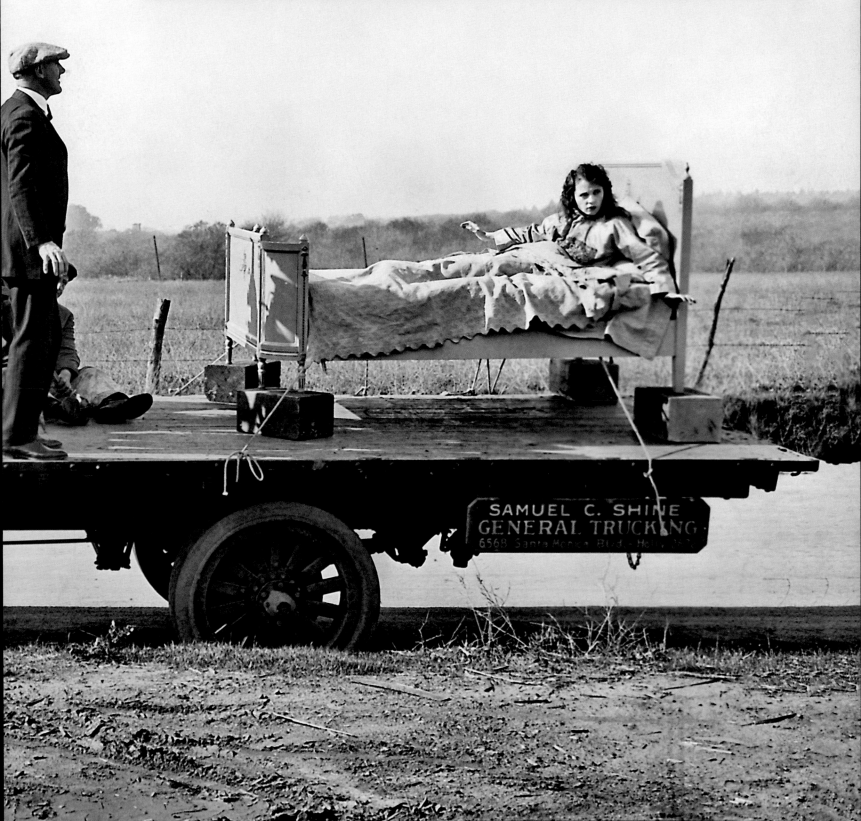

SAMUEL C. SHINE
GENERAL TRUCKING
6568 Santa Monica Blvd. Holly.

# Streamlining Comes to Trucks

The 1930s Art Deco movement brought streamlining to the truck world, from big vehicles operated by major brands to light commercial trucks run by small businesses. These sleek, new bodies, whether on car-derived pickups or purpose-built tankers or haulers, were part of a conscious attempt by manufacturers to make their trucks attractive as well as functional. Trucks are now not just workhorses – they have desirability, imbuing drivers with a new sense of pride. The 1930s was one of the most bright and beautiful times for trucks.

### △ Diamond T406 Deluxe

| | | |
|---|---|---|
| **Date** 1934 | **Origin** USA | |
| **Engine** 6-cylinder petrol, 90 hp | | |
| **Payload** 1.5 tonnes (1.5 tons) | | |

With its Packard dashboard and interior, driving this Diamond T felt akin to driving a luxury car. The "Express Coachwork" rear enabled goods to be placed in the cargo area without the risk of items falling out if not fully secured.

### ▷ Dodge K34 Tanker

**Date** 1934  **Origin** USA

**Engine** 6-cylinder petrol, 95 hp

**Payload** 2 tonnes (2 tons)

In the 1930s, streamlining tanker trucks with sleek, flowing bodies became a modern advertising statement. This K34 kept the weight of its contents over the chassis while the sides of the bodywork were used for storage and featured access steps to fill the tanker. Its pumping equipment was fully enclosed at the rear.

Hood
mascot

Photo courtesy of Hyman Ltd. www.hymanltd.com

Roll-down
windows

### ◁ Diamond T 80 Deluxe

**Date** 1936  **Origin** USA

**Engine** 6-cylinder petrol, 73 hp

**Payload** 680 kg (1,500 lb)

Made for just two years, the 80 Deluxe had an attractive style with a long cargo bed and Hercules DX engine. Diamond Ts had a distinctive design with split-screen cabs and flared fenders.

Custom-made
aerofoil skin
covers tanker

### ▽ Mercedes W136 – 170V

**Date** 1937  **Origin** Germany

**Engine** 4-cylinder petrol, 34 hp

**Payload** 340 kg (750 lb)

Introduced in 1937, the 170V car became Mercedes' most affordable model. It featured a strong tubular chassis that formed the basis of both van and pickup versions. Rather than boxy panels, these van bodies had contemporary style.

DISKONTO-BANK
DISKONTO GESELLSCHAFT
Wittenbergplatz · Unter den Linden 33 · Berlin

Coach-built
van body

Flat-sided rear
pickup panels saves
manufacturing costs

△ **Studebaker Coupe Express**

| | |
|---|---|
| **Date** 1937 | **Origin** USA |

**Engine** 6-cylinder petrol, 86 hp

**Payload** 500 kg (1,102 lb)

This pickup was based on the Studebaker
Dictator. While it retained the cars' styling,
it featured a practical load area at the rear
and created a new market for stylish utility
vehicles that was followed by Ford and GM.

△ **Peugeot 202 Pickup**

| | |
|---|---|
| **Date** 1938 | **Origin** France |

**Engine** 4-cylinder petrol, 23 hp

**Payload** 350 kg (772 lb)

The heavily Art Deco-inspired Peugeot 202
was an affordable, stylish pickup. The lights
positioned behind the grille enhanced its
aerodynamic shape and would also have
protected them from stone chips.

Snub-nose cab
inspires a new
range of trucks

Photo courtesy of Hyman Ltd. www.hymanltd.com

△ **Chevrolet Coupe Pickup**

| | |
|---|---|
| **Date** 1938 | **Origin** USA |

**Engine** 6-cylinder petrol, 85 hp

**Payload** 150 kg (331 lb)

In an exercise of style over substance,
Chevrolet's pickup offered no
compromises in speed or comfort.
However, the cargo area was simply
an extension of the boot.

△ **Ford COE**

| | |
|---|---|
| **Date** 1940 | **Origin** USA |

**Engine** V8 petrol, 85 hp

**Payload** 1–1.5 tonnes (1–1.5 tons)

The popular COE (cab over engine) range
offered a sense of comic book futurism. The snub
nose brought the driving area forward, increasing
the length available for loading on the chassis.

△ **Dodge RX70 Airflow**

| | |
|---|---|
| **Date** 1940 | **Origin** USA |

**Engine** 6-cylinder petrol, 100 hp

**Payload** 3 tonnes (3 tons)

Inspired by the Chrysler Airflow car, these
trucks reflected the vibrancy of the times.
Their sleek appearance was popular with
oil companies keen to portray a modern
image – just 303 were built.

# Going the Distance

Across the desolate Australian outback, through war-torn regions in the Middle East, along dramatic American coastal highways, or on transcontinental treks to China – some truck journeys are truly epic. Whether it is the vast distances, spectacular scenery, or nerve-shredding danger that sets such routes apart, what unites them all is the fundamental need to deliver intact cargo to its destination – wherever in the world.

Trucks are built to travel long distances. Every day around the world, thousands of trucks criss-cross countries, even continents, to deliver goods to their destinations. Truck journeys are mostly plain sailing – modern roads are often smooth, and motorways and highways allow drivers to maintain steady speeds. But some trips are longer and more difficult than others, even hazardous – they call for brave drivers.

## Road trains

Australia is renowned for its vast open spaces of bush and outback, separating major towns and cities, which means it has some of the most epic truck routes. For example, the distance from the north to south coasts, between Darwin and Adelaide, is around 3,000 km (1,864 miles).

Drivers of road trains – tractor units pulling a multitude of trailers with as many as 100 wheels – can travel for hours without seeing another person.

The monster-sized road trains are possible only because the roads they travel on are so straight and, like much of Australia, so sparsely populated. In a country served only by a limited railway network, road trains have become the most effective and economic way of transporting livestock, fuel, and other bulk goods across such challenging distances. Just because the road is straight, and

**Lifeline in the outback**
A road train powers along the Great Northern Highway in the desolate Pilbara region of Western Australia. Overtaking such a long truck is a challenge, and so is the temperature, which can reach more than 49 °C (120 °F).

the scenery often spectacular, does not mean the driving is easy. The heat is often extreme, and if a storm hits, roads can quickly turn to mud. Also, when a road train breaks down, there is unlikely to be a recovery service nearby. So drivers are likely to have to fix problems themselves.

## Driving into danger

Other long-distance truck routes present different kinds of challenge – some that can even threaten a driver's life. In the 2010s, one of the most dangerous journeys was from

the border of Jordan into western Iraq. Since 2014, this has meant passing through territory controlled by the radical group Islamic State of Iraq and Syria (ISIS).

People in Iraq were desperate for supplies – such as food and medicine – and a brave group of drivers were there to deliver them. To get into Iraq, drivers had to negotiate long stretches of lawless roads, facing the ever-present threats from local militias and ISIS roadblocks, and paying "tolls" before they could continue. With

---

**TALKING POINT**

## Autonomous Truck World Record

While a journey of just over 212 km (132 miles) might not seem epic, it is when the vehicle involved has no driver. In October 2016, an autonomous – self-driving – Volvo truck travelled from Fort Collins to Colorado Springs in Colorado, US, without input from the human driver or the assistance of a lead vehicle. For much of the trip, – a world record for an autonomous truck – the driver onboard was able to watch the road from the comfort of the truck's sleeper berth. The journey was also claimed as the first commercial shipment by a driverless truck. More than 51,000 cans of beer safely reached their destination.

**A program engineer involved** in the development of autonomous vehicles, keeps an eye on the road – the only human intervention on the driverless truck's journey.

ROAD TRAIN ROUTE
MAX 53.5 m LONG
TAKE CARE WHEN OVERTAKING

> ## "**Every day**, you wake up **somewhere different**. You have **sunrises and sunsets**. It's the same sun, but it's **different everywhere**."

SUSAN ZIMMERMAN, AMERICAN TRUCK DRIVER

smugglers to contend with as well, drivers constantly faced the prospect of being robbed.

What had once been a prosperous route, travelled by thousands of trucks every day in the 1990s, was now a driver's nightmare, with the number of drivers dropping to the hundreds – and even tens – during the worst of the fighting. Yet the most courageous drivers still made the trips, often driving in convoys for safety. Money was a major motivator – some drivers could triple their usual wages by risking runs into the area.

## Crossing continents

Not all epic drives are dangerous – some are just incredibly long. One of the longest runs from Germany, across Eastern Europe and Asia, to northwestern China and the city of Ürümqi – capital of the Xinjiang region. Covering about 8,000 km (5,000 miles) the journey can take more than two weeks, and crosses numerous countries, including Poland, Belarus, Kazakhstan, and Russia. In 2018, an even longer route, the Western Europe–Western China Highway, was completed, stretching 8,445 km (5,250 miles) from St Petersburg, on Russia's Baltic Sea coast, to the eastern Chinese port of Lianyungang. A truck can cover this route in as little as 10 days – carrying freight more quickly than by sea, or even rail, and more cheaply than by air.

## Taking in the view

Long-distance routes can involve seemingly endless stretches of featureless country. Some truck drivers, however, get to enjoy more picturesque journeys. In the United States, some routes are renowned for their amazing scenery. Interstate 70 (I-70) Highway spans 3,465 km (2,153 miles) through ten US states, including Utah and Colorado. Drivers can witness deserts, deep canyons, steep cliffs, and dramatic rock formations.

In California, State Route 1 (SR 1), also known as the Pacific Coast Highway, runs for 1,055 km (656 miles) from north of San Francisco to south

**Scenic but demanding**
Desert rock formations are a majestic sight on the I-70 Highway in Utah. Its remoteness can bring danger – one stretch of over 160 km (100 miles) has no exits or services whatsoever.

of Los Angeles. It offers drivers magnificent coastal views, striking mountain and forest backdrops, and man-made wonders, such as The Golden Gate Bridge in San Francisco and Bixby Bridge, south of Monterey.

Being out on the open road can be monotonous, physically draining, and full of risk. However, experiencing places and sights that few ever see is rich reward for many truck drivers.

# Emergency Vehicles

As urban communities and industrial centres continued to expand in the 1920s and 1930s, so the need for increasingly sophisticated and adaptable trucks to serve as emergency vehicles increased. The most common example during this period was the unmistakably red fire engine. By their nature, most fire-fighting trucks were big and heavy, carrying ladders, hoses, and other equipment, and large quantities of water that requires powerful pumps. As chassis design developed, and engines became more efficient and refined, many smaller trucks were converted into other emergency vehicles, such as ambulances and police vans. As mainstream manufacturers entered the truck market, emergency services could order entire fleets of vehicles to meet their needs.

### ▽ Leyland Fire Engine

| | |
|---|---|
| **Date** 1920 | **Origin** UK |

**Engine** 4-cylinder petrol, 55 hp

**Payload** 3 tonnes (3 tons)

Although the design of the Leyland Fire Engine, with its solid tyres and cast metal wheels, dated back to World War I, it was widely used in towns and cities throughout the UK in the 1920s. A shaft – known as the power take-off – linked the engine at the front to the water pump at the rear.

**Escape and rescue** ladder on sliding rollers

**Water pump** powered by the engine

FIRE BRIGADE 1920

**Water suction** pipe to clear and flush drains and gullies

### ▽ Ahrens-Fox K-S-4 Fire Engine

| | |
|---|---|
| **Date** 1922 | **Origin** USA |

**Engine** 6-cylinder petrol

**Payload** 2.7 tonnes (2.7 tons)

Fire engines made by Ahrens-Fox were easily identified by the chrome pressured sphere above the pump. This ensured the delivery of a constant high water pressure from the pump to the hose nozzle.

**Throttle control** on steering column

**Chemical tank** holding soda acid – added to water as a fire extinguisher

### △ Ford Model T Fire Engine

| | |
|---|---|
| **Date** 1923 | **Origin** USA |

**Engine** 4-cylinder petrol, 26 hp

**Payload** 340 kg (750 lb)

The Model T truck chassis underpinned many types of vehicle, including this fire truck. Being relatively small, it was nimble and manouverable. Ford Model T fire engines were used in the US, Europe, and Australia.

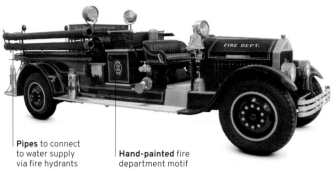

**Pipes** to connect to water supply via fire hydrants

**Hand-painted** fire department motif

◁ **Pierce-Arrow Fire Engine**

Date 1924  Origin USA

Engine 4-cylinder petrol

Payload 2.7 tonnes (2.7 tons)

Best known for its large, powerful luxury cars, Pierce-Arrow also produced trucks from 1911. As well as buses and goods vans, the company adapted its truck platforms into medium-size fire engines.

▷ **Volvo PV650 Ambulance**

Date 1929  Origin Sweden

Engine 6-cylinder petrol, 55 hp

Payload 1.5 tonnes (1.5 tons)

Volvo produced its first car in 1927 and first truck a year later, at a time when the Swedish market was dominated by American manufacturers. The large PV650 chassis was used for luxury car platforms but also proved a sturdy and adaptable base for ambulances, light trucks, and even hearses.

▽ **Ford AA Patrol Van**

Date 1930  Origin USA

Engine 4-cylinder petrol, 40 hp

Payload Variable

The Ford AA chassis served as the base for many types of truck, including police patrol vans, used to transport suspects. The van's siren sits on the front fender.

△ **Ford Police Ambulance**

Date 1935  Origin USA

Engine V8 flathead petrol, 85 hp

Payload 0.9 tonnes (0.9 tons)

Across the US, police forces specified Ford V8s in many of their vehicles. This ambulance, with a specialist coach-built body, began service in 1935 in Endicott, New York State, in response to the growing number of accidents involving automobiles. The small side door gave officers access to the back.

**Fire hose** reel drum

**Two-tier** wooden escape ladder

◁ **Leyland Lynx Fire Engine**

Date 1939  Origin UK

Engine 6-cylinder petrol

Payload 3 tonnes (3 tons)

The Leyland factory, like many others, had its own fire department, using its own trucks as fire engines, including the Lynx. A roof was not necessary as the truck only responded to calls within the grounds of the factory.

# Ahrens-Fox N-S-4

The Ahrens-Fox Fire Engine Company was founded in Ohio, USA, in 1910. Until then most fire tenders had been cumbersome horse-drawn wagons, but the advent of internal combustion meant large, engined trucks quickly became popular. While other fire trucks of the time, such as the Ward LaFrance, shared a similar conventional design, the Ahrens-Fox had a distinctive appearance with its pump set in front of the radiator grille over the front axle.

**A COMBINATION OF PUMP AND HOSE CAR,** the Ahrens-Fox N-S-4 was highly regarded. While most fire trucks built by independent brands used existing engines, Ahrens-Fox built its own, making this one of the first fire trucks designed and manufactured from the ground up. Agile enough to get to the job faster than larger vehicles, it sported a chromed pressurized sphere above the pump that delivered water at constant high pressure and prevented surges caused by sudden pressure losses. The truck could eject up to 5,000 litres (1,100 gallons) of water per minute – revolutionary for its time. Fire departments took great pride in keeping their Ahrens-Fox vehicles immaculate – upon returning to the garage, cleaning and polishing would begin just as soon as the fuel tank for the 16-litre engine had been replenished.

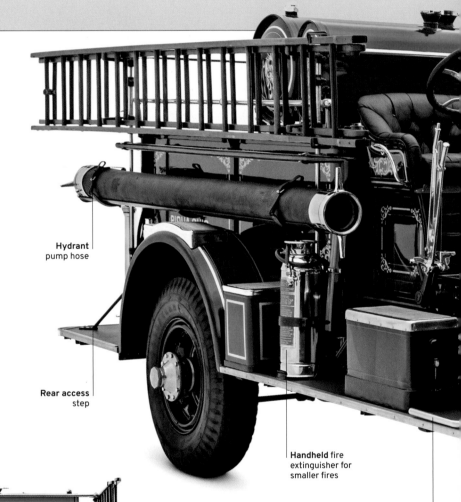

Hydrant pump hose

Rear access step

Handheld fire extinguisher for smaller fires

Engine starting handle

| SPECIFICATIONS | |
|---|---|
| Model | Ahrens-Fox N-S-4 |
| Origin | USA |
| Assembly | Ohio |
| Production run | Unknown |
| Weight | Unknown |
| Payload | 1.8 tonnes (1.8 tons) |
| Engine | 6-cylinder petrol, 16,354 cc, 110 bhp |
| Transmission | 3-speed manual |
| Maximum speed | 80 km/h (50 mph) |

**FRONT VIEW**

**REAR VIEW**

**All in the name**
Every Ahrens-Fox vehicle has a maker's plate attached on the bodywork. The company logo along with the factory location is stated. Although most Ahrens-Fox engines were sold within the US, having the factory address could be useful for overseas enquiries and for obtaining spares.

## THE DETAILS

**1.** Six-cylinder engine, two cylinders per block **2.** Presentation plate for the original fire department **3.** High-pressure chrome water nozzles **4.** Siren **5.** Water pump showing the fittings to connect to hydrant

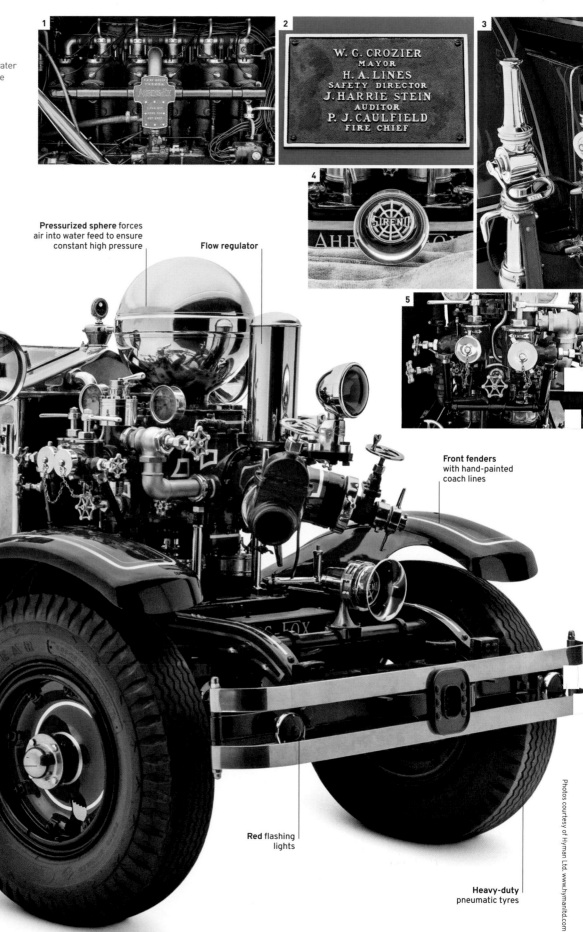

W. G. CROZIER
MAYOR
H. A. LINES
SAFETY DIRECTOR
J. HARRIE STEIN
AUDITOR
P. J. CAULFIELD
FIRE CHIEF

**Headlamp** fixed to the radiator

**Pressurized sphere** forces air into water feed to ensure constant high pressure

**Flow regulator**

**Front fenders** with hand-painted coach lines

**Wheel hub** cover plate

**Red** flashing lights

**Heavy-duty** pneumatic tyres

### Innovative design

It may seem strange to fit the pump in front of the radiator, which needs air flowing around it – especially given the pace at which a fire truck like this would have been driven. But the front of the truck was also the closest to the scene upon arrival, making all the hose valves and apparatus instantly accessible.

# 1939–1959
# IN WAR AND PEACE

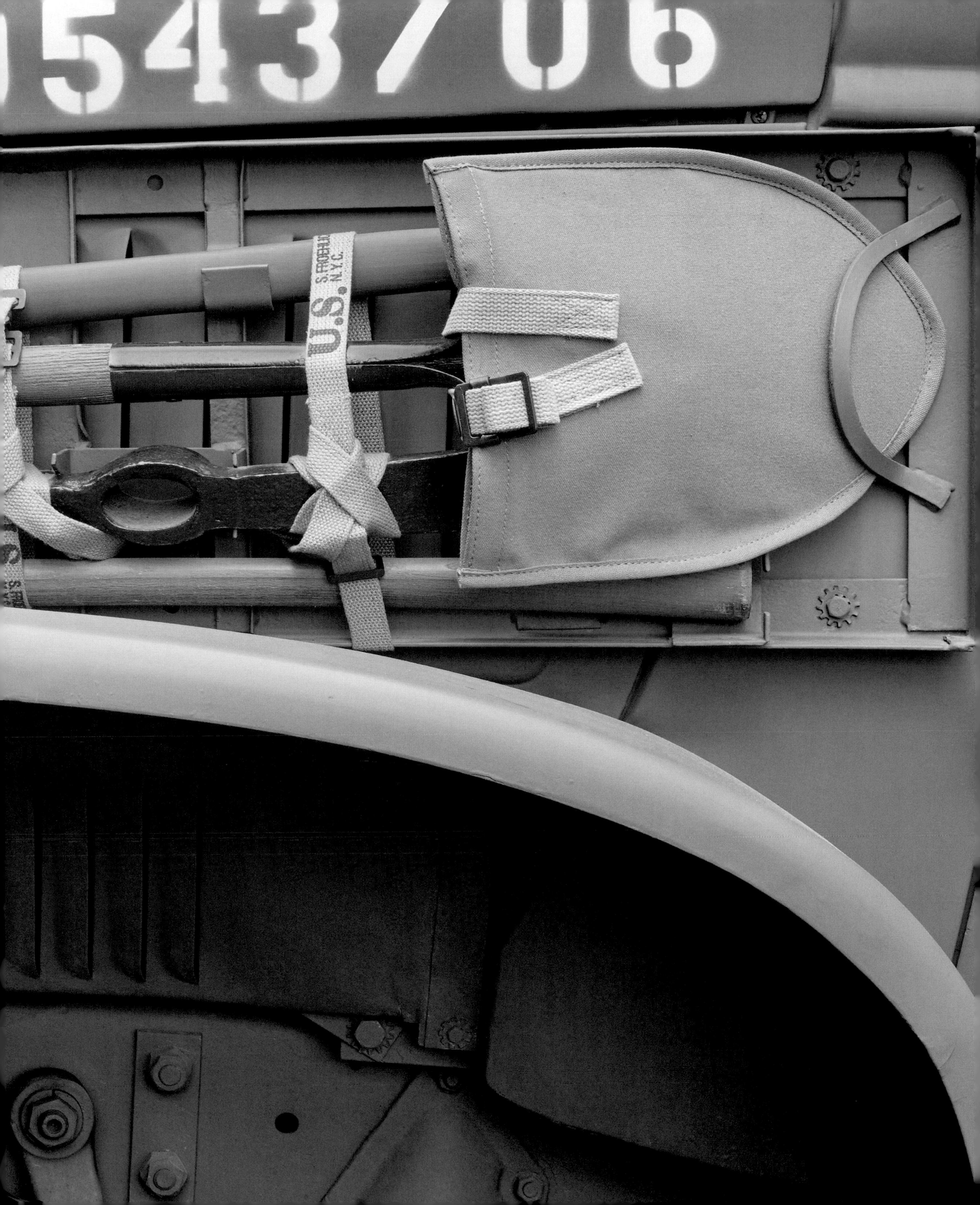
543706
U.S.
N.Y.C.

# IN WAR AND PEACE

**The outbreak of World War II** created a massive demand for trucks around the globe. Between 1941 and 1945, the US – which had significantly expanded its industrial infrastructure – supplied the Allied forces with millions of tonnes of trucks under the Lend-Lease agreement, a system that allowed the US to distribute war supplies to its allies.

During the postwar recovery, mass auctions were held across Europe to sell the resulting surplus military vehicles. The French fire service was one of the largest purchasers of these, and the six-wheel-drive GMC CCKW started to play its part in saving lives across France, with many remaining in service until the 1980s.

In the 1950s, the American dream was in full swing, and there was a widespread sense of liberty. Many wartime truck drivers returned home and began serving civilian businesses, while at the same time, the demand for more comfortable and refined trucks grew. China, Korea, and parts of Eastern Europe were dependent on Soviet trucks, while India and many other nations imported British models.

Truck manufacturers across the world improved their engines and functionality, and brand loyalty started to develop. When the Chevrolet 3100 series started to compete against Ford's F-series in the US, it sparked devoted fans of each. No longer seen as merely utility vehicles, trucks became a lifestyle choice for many, expanding "trucking culture" to a wider audience.

△ **1955 Tempo advertisement**
German manufacturer Tempo produced military trucks during World War II, then moved to civilian production following the war.

△ **Unusual load**
Trucks now had the power to haul large and unexpected loads. In 1945, this entire house was moved on a flatbed truck in Wisconsin.

> "**Ambition** is a dream with a V8 engine. Ain't **nowhere else** in the world where you can go from **driving a truck** to a Cadillac overnight."
>
> ELVIS PRESLEY, WHEN HE WAS A TRUCK DRIVER

## Key events

▷ **1944** GMC, Diamond T, Ward La France, and others are produced with open cabs to save on resources.

▷ **1944** By the end of the day on 6 June, over 170,000 vehicles, mostly trucks, have reached France as part of the D-Day landings in Normandy.

▷ **1946** Czechoslovakian company Praga produces the A150 light civilian truck.

▷ **1946** Russian manufacturer, GAZ, introduces the GAZ-51 light truck.

▷ **1947** Due to oil shortages, Datsun experiments with electrically powered light pickup trucks in Japan.

▷ **1948** Ford introduces its new F-series range of small pickup trucks.

△ **Debut of an iconic truck**
Ford's introduction of a larger, newly designed medium duty F-series created a sales boom among agricultural and industrial customers.

▷ **1949** DAF produces its first truck, the A30, in the Netherlands.

▷ **1952** The Fiat 652 truck enters production. Its distinct "smiling face" grille is featured in movies and is adopted by the Ferrari race team.

▷ **1956** The first purpose-built container ship is launched in British Columbia, Canada, creating the need for trucks to transport containers by road.

▷ **1956** Australian road trains largely retire the under-powered British Leyland Beavers in favour of new American Macks and Kenworths.

◁ **1949 advertisement for Panhard,** a French truck manufacturer.

# War Machines

The demand created by World War II for military vehicles, in particular trucks to mobilize the forces, exceeded levels ever seen before. Rapid developments and continuous production enabled factories to keep producing standarized models ready for deployment. Trucks, like those pictured here, lined streets in the UK during the secret build-up to the D-Day landings, then mysteriously "disappeared" overnight.

## ALLIED ASSEMBLY LINES
In the US, truck companies were operating beyond full capacity to supply, not only vehicles for their own nation's efforts, but also for the recently established Lend-Lease Agreement that allowed them to supply assets to assist the war effort overseas. Thousands of trucks were supplied to Allied nations, shipped around the world to the USSR, Australia, and the UK. Many were delivered as flat-pack kits to reduce assembly time and enable greater shipping volumes. These kits ranged from Jeeps to Diamond T 981 tank transporters, which could be assembled with basic tools in a very short time. In Australia, large quantities of Studebaker trucks were dispatched, with over half of the 197,000 produced going to the USSR.

**Bedford MW and OY trucks** commissioned by the British Army await deployment during D-Day outside the Vauxhall Factory in Luton, UK.

# Divco-Twin Model U

Founded in 1926, Divco – Detroit Industrial Vehicle Company – specialized in building small vans. After merging with Twin Coach Co. in 1936, the Model U was launched the following year, primarily designed for the doorstep delivery of dairy products. Although Divco remained a relatively small-scale manufacturer, by Detroit standards, the Model U became a familiar sight in US cities, remaining in production for almost five decades.

**THE DESIGN OF THE DIVCO-TWIN MODEL U** centred on ease of use for the delivery driver, and featured clever innovations throughout. With the accelerator and brake controls mounted on the steering column, it could be driven either seated or standing up, which allowed the driver to easily and quickly hop in and out while making deliveries – as did it's semi-automatic cantilevered folding doors. While early models were merely insulated to keep the products fresh, refrigerated versions were developed in the 1950s. Such was success of the overall design, that with only minor updates, the van remained in production until 1986, when the company finally ceased trading.

Welded, steel bodywork

Rear fender

Cantilevered folding door

Heavy duty commercial tyres

**FRONT VIEW**

**REAR VIEW**

**Divco's legacy**
While the Divco name remained a constant, like many manufacturers, ownership of the company changed many times over its near sixty-year history.

| SPECIFICATIONS | |
| --- | --- |
| Model | Divco-Twin Model U |
| Origin | USA |
| Assembly | Detroit |
| Production run | 1937–86 |
| Weight | 4.1 tonnes (4.5 tons) |
| Payload | 1.4 tonnes (1.5 tons) |
| Engine | 4-cylinder petrol Continental, 2,290 cc, 38 hp |
| Trasmission | 4-speed manual |
| Maximum speed | 64 km/h (40 mph) |

1

2

3

4

5

6

**Windscreen**
wiper

**Driving mirrors**
are aimed at the
kerb and the side

**Divco grille**
badge

## THE DETAILS

**1.** Continental engine – Continental Motors acquired Divco in 1932 **2.** Top view of engine showing calling siren and horn **3.** Cabin heater **4.** Column-mounted controls **5.** Dashboard instrumentation **6.** Swing out driver's seat for quick access and exit

**Distinctive snub nose**
became recognizable
design feature

**Headlamps**

### Versatile vehicle
Divco vans featured in the background of urban scenes in many old Hollywood movies. More recently, derelict vehicles have been converted into custom lowriders and drag racers. Others have been restored to their original glory, such as this example.

# Trucks Behind the Iron Curtain

After World War II, the division of Europe between the democratic west and the communist east – separated by the so-called "Iron Curtain" – disrupted many motor factories. Under communist rule, particularly in East Germany and the USSR, new companies were hastily formed to manufacture trucks – largely, at first, for the police and military. Advances in production were slow, but vehicles were built in huge quantities. Heavy, aggressive-looking machines rolled into cities and army camps, while new medium-sized trucks and car-based vans met civilian needs.

### ▷ GAZ-51

**Date** 1946 **Origin** USSR

**Engine** 6-cylinder petrol, 68 hp

**Payload** 2 tonnes (2 tons)

The Studebaker US6, provided by the US to Soviet forces during World War II, inspired the 4×4 GAZ-51. Such was the demand for the truck that production continued until 1975. The GAZ-51 was also built under licence in North Korea as the Sungri 58 and in China as the Yuejin NJ-130.

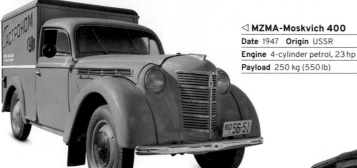

### ◁ MZMA-Moskvich 400

**Date** 1947 **Origin** USSR

**Engine** 4-cylinder petrol, 23 hp

**Payload** 250 kg (550 lb)

The box-van 400, like all Moskvich models, was intended to provide affordable mass transport. Based on the German pre-World War II Opel Kadett car, it was produced by MZMA (Moscow Small Car Factory).

### ▷ IFA G5

**Date** 1952 **Origin** East Germany

**Engine** 6-cylinder diesel, 120 hp

**Payload** 5 tonnes (5 tons)

The triple-axle, six-wheel G5 was popular with East German fire services, as well as the military and police. It was first developed during World War II, then produced by IFA – a nationalized union of vehicle manufacturers founded in 1948 under Soviet supervision.

### ▽ GAZ-69

**Date** 1953 **Origin** USSR

**Engine** 4-cylinder petrol, 55 hp

**Payload** 500 kg (1,100 lb)

The Soviet answer to the Jeep, the GAZ-69 light utility truck was produced for 30 years and exported around the world. Its 4×4 off-road agility made it a mainstay of the Soviet Army in command car, pickup, and van versions.

**Roof-mounted** high-pressure water cannon

**Water pump** was driven from the engine

**Marker** to indicate front of vehicle to driver

**Curved body lines** aimed at the American market

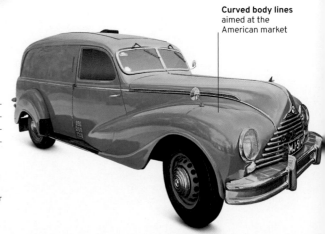

### ▷ EMW-340 van

**Date** 1954 **Origin** East Germany

**Engine** 6-cylinder petrol, 54 hp

**Payload** 250 kg (550 lb)

The sleek lines of the EMW-340 van mirror those of the BMW-340 saloon, renamed as EMW in 1952, following the postwar split of the two BMW factories between East and West Germany. Fewer than 1,000 were made.

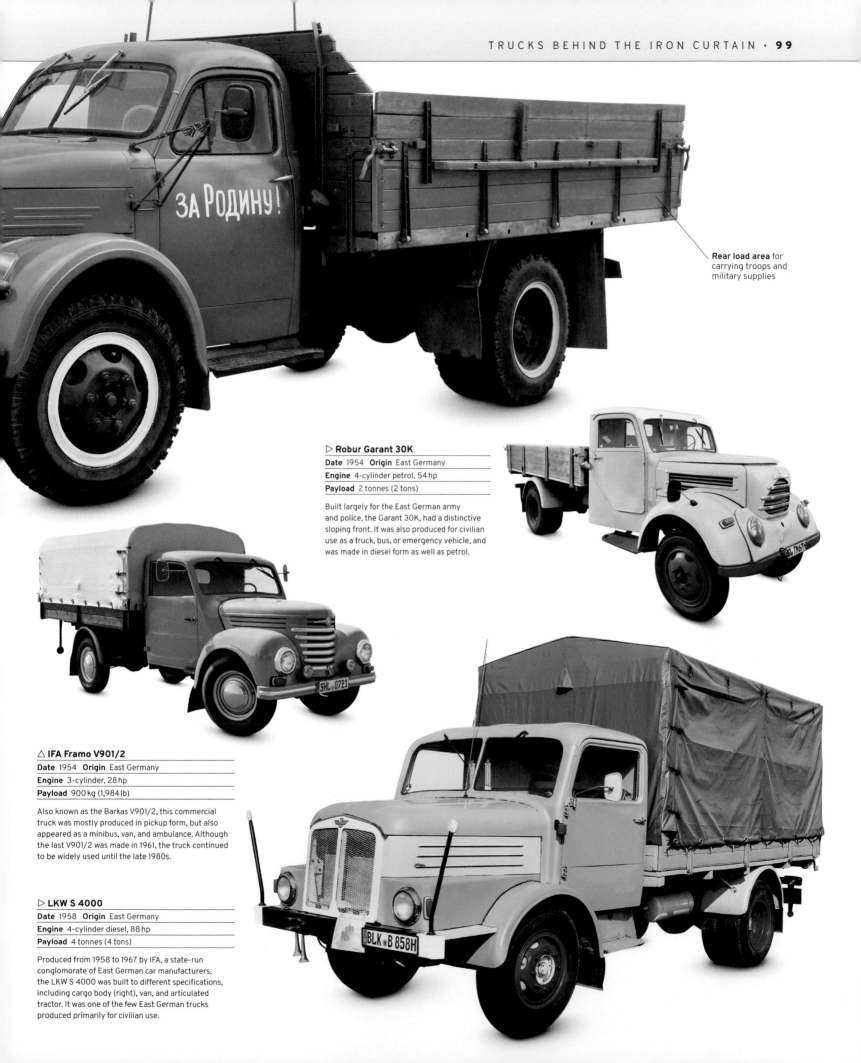

ЗА РОДИНУ!

**Rear load area** for carrying troops and military supplies

▷ **Robur Garant 30K**

**Date** 1954  **Origin** East Germany

**Engine** 4-cylinder petrol, 54 hp

**Payload** 2 tonnes (2 tons)

Built largely for the East German army and police, the Garant 30K, had a distinctive sloping front. It was also produced for civilian use as a truck, bus, or emergency vehicle, and was made in diesel form as well as petrol.

△ **IFA Framo V901/2**

**Date** 1954  **Origin** East Germany

**Engine** 3-cylinder, 28 hp

**Payload** 900 kg (1,984 lb)

Also known as the Barkas V901/2, this commercial truck was mostly produced in pickup form, but also appeared as a minibus, van, and ambulance. Although the last V901/2 was made in 1961, the truck continued to be widely used until the late 1980s.

▷ **LKW S 4000**

**Date** 1958  **Origin** East Germany

**Engine** 4-cylinder diesel, 88 hp

**Payload** 4 tonnes (4 tons)

Produced from 1958 to 1967 by IFA, a state-run conglomorate of East German car manufacturers, the LKW S 4000 was built to different specifications, including cargo body (right), van, and articulated tractor. It was one of the few East German trucks produced primarily for civilian use.

# Heavy Haulage in the 1950s

With the rise of consumerism, mass production, and infrastructure development after World War II, bigger and more powerful trucks were required to move the larger volumes necessary to support the economic boom. During the 1950s, V8s and diesel engines improved torque and fuel efficiency, while hydraulic brakes became more widely used, allowing heavier loads to be carried more safely. Driver comfort also improved dramatically thanks to the adoption of automatic transmission and power steering, reducing driver fatigue on long-haul journeys.

Mirror

Fuel tank

### ▽ Corbitt 22TG 4x2

**Date** 1949   **Origin** USA

**Engine** 6-cylinder diesel or petrol 127 hp

**Payload** 4.5 tonnes (4.4 tons)

The Continental flat-head model B-6427 engine used in this truck was criticised for not having enough power. However, it was a well-built truck with a Fuller transmission, Timken axels, Westinghouse air brakes, and Parish Chrome Manganese heat-treated steel frames.

### △ Mercedes L6600

**Date** 1950   **Origin** Germany

**Engine** 6-cylinder diesel, 145 hp

**Payload** 6.6 tonnes (6.5 tons)

Mercedes-Benz was renowned for engineering excellence and durability. Classified as a medium- to heavy-duty truck, the L6600 took its name from its payload capacity and was regarded as "the truck that rebuilt Germany".

**Front-opening** windscreen

**Fifth wheel** for attaching trailers

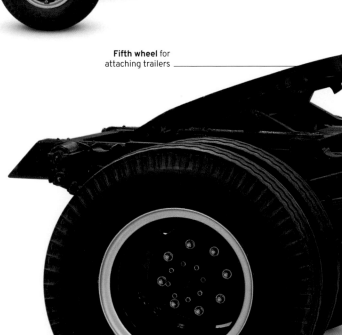

### △ Ford F-7 4x2

**Date** 1950   **Origin** USA

**Engine** Flathead V8, 145 hp

**Payload** 9–14 tonnes (8.9–13.4 tons)

The largest of the Ford F-series, the F-7 "Big Job" came with enough power to haul truly heavy loads. But the compact design of the engine, in which the exhaust passed through the "V" between the two cylinder banks, made it prone to overheating.

◁ **White Superpower 4x2**

**Date** 1950s **Origin** USA

**Engine** Straight 6 petrol, 215 hp

**Payload** 9–14 tonnes
(8.9–13.4 tons)

A veteran engineering firm that had built steams trucks years before, White made everything from cars to sewing machines. After World War II, it dedicated itself exclusively to producing heavy trucks.

**Storage** areas

▽ **Diamond T 950RS 6x4**

**Date** 1952 **Origin** USA

**Engine** 6-cylinder diesel, 330 hp

**Payload** 15.4–18 tonnes (15–17.9 tons)

Like many other truck manufacturers, Diamond T earned a reputation for making rugged, powerful, heavy trucks for the military during the war. Its experience building tank transporters meant it was ideally placed to manufacture heavy haulage trucks.

**Dual** rear axle provides greater weight capacity than a single axle

**V12** engine

◁ **Berliet T100**

**Date** 1957 **Origin** France

**Engine** 29.6-litre V12 diesel, 700 hp

**Payload** 50 tonnes (49 tons)

At 5 metres (16.4 feet) wide, this behemoth was designed for off-road use in the oil and mining industries. With a gross weight of 100 tonnes (98 tons), the T100 was the largest truck in the world at the time.

▽ **Mack H63T 4x2**

**Date** 1958 **Origin** USA

**Engine** Petrol and diesel V8, 170 hp

**Payload** 9–13.6 tonnes (8.9–13.4 tons)

Despite having an integrated sleeping compartment, the H63T's "cab-over" layout allowed the tractor unit to be short, maximizing the trailer length. The H-series was nicknamed "Cherry Picker" because of its high cab.

▽ **Mack G73**

**Date** 1959 **Origin** USA

**Engine** Mack and Cummins, up to 335 hp

**Payload** 15.4–18.1 tonnes
(15.2–17.9 tons)

Mack's first aluminium-riveted cab-over-engine truck featured an innovative door panel on the left front of the cab to allow easy access to the electrical wiring and air lines.

# French Trucks

Since the dawn of the motorcar, France has pioneered with trucks, or *camions*. The French established the world's first motor races, and the early brands that competed in them went on to become some of Europe's leading truck manufacturers. France had a classification system called *Voider Sans Permis* for small vehicles that weighed under 425 kg (937 lb) or had moped engines. These required no licence to drive, giving rise to a quirky range of smaller trade vehicles. In the interests of mobilizing France again after WWII, tiny delivery vehicles were adapted from such designs, some of which even had their origins in pedal-powered vehicles. Meanwhile, leading manufacturers were developing a new era of medium and heavy truck chassis. Makeshift roads that had been rapidly carved across the nation during the war as cargo routes were also being improved to create one of the best examples of road transport and haulage systems in the world.

**Wood-panelled** cargo body

▷ **Peugeot Q3A**

| Date | 1948 | Origin | France |
|---|---|---|---|

**Engine** Petrol, 40 hp

**Payload** 1.4 tonnes (1.4 tons)

Peugeot introduced the DMA, its first cab-over, in 1941, when the company was under German control. The model was only available in grey during the war, using paint left over from the German army. Afterwards, it was renamed the Q3.

**Small motorcycle** headlamps

▷ **Mochet CM125**

| Date | 1952 | Origin | France |
|---|---|---|---|

**Engine** Single-cylinder, 4 hp

**Payload** 100 kg (220 lb)

Before WWII, Mochet was known for pedal-powered velocars, some with pickup bodies. Postwar, these were fitted with a small engine with a pull-start under the dashboard. This vehicle was the cheapest four-wheeled truck of its day, and had a top speed of 35 km/h (22 mph).

**TALKING POINT**

## Berliet

Although Berliet began making trucks before World War I, its wartime production boosted the company's success and paved the way for it to become one of France's leading truck makers. Initially known for building luxury cars, Berliet began truck building in 1906, producing models for military and civilian purposes. By the 1950s it was a leader in on- and off-road trucks in France.

**1955 Berliet advertisement** Berliet lead the development of heavy articulated semi trucks during the 1950s, which were becoming increasingly popular across many industries.

△ **Peugeot D3**

| Date | 1954 | Origin | France |
|---|---|---|---|

**Engine** Petrol, 30 hp

**Payload** Unknown

Chenard-Walcker originally produced the D4 before Peugeot acquired the French company. Fitted with a Peugeot 203 car engine, this van – a direct rival to Citroën's H van – was very basic but simple to maintain.

**Operator's livery**

**Radiator grille**

▷ **Unic Zu100**

| Date | 1954 | Origin | France |
|---|---|---|---|

**Engine** 6-cylinder diesel, 135 hp

**Payload** 3 tonnes (3 tons)

The Unic Zu100 had a very long bonnet and used an advanced 8-speed air-operated gearbox. These trucks were popular around Europe, with uses ranging from removal and livestock vehicles, to tankers.

▷ **Mochet CM-125Y**

**Date** 1956 **Origin** France

**Engine** Single-cylinder, 4 hp

**Payload** 100 kg (220 lb)

The CM-125Y used the same running gear as the car it replaced but with an updated van-like body. Some, such as this one, used metal cargo bodies rather than canvas, which did nothing to help its slow speed.

▽ **Somua JL 19**

**Datec** 1957 **Origin** France

**Engine** 6-cylinder diesel, 180 hp

**Payload** 26 tonnes (25.6 tons)

The JL19 was produced by a prolific French machinery and vehicle manufacturer, known for making beautiful Parisian buses. It had varied chassis options; tractor units such as this one were often used as airport tankers, while others had large van bodies.

**Large door mirrors** offer increased rear visibility

**Cab roof** has storage space for extra cargo and supplies

△ **Berliet GLR**

**Date** 1958 **Origin** France

**Engine** 5-cylinder diesel, 125 hp

**Payload** 5 tonnes (4.9 tons)

The GLR was a heavy truck, with a generous payload for its size. The open-back Berliet pictured is typical of the body types used on this chassis and has a 5-speed gearbox.

▷ **Berliet GBC-8 Gazelle**

**Date** 1959 **Origin** France

**Engine** 5-cylinder diesel, 125 hp

**Payload** 4 tonnes (3.9 tons)

The 6x6 GBC was developed for export to the North African market. With its off-road capabilities and robust build, it could cope with extreme desert conditions, while performing vital logistics roles.

# Peterbilt 281

In 1939, Theodore Alfred "Al" Peterman bought the Fageol truck company of Oakland, California. Prior to this, he had been in the lumber business, purchasing old army trucks and adapting them to haul logs. The first Peterbilt trucks were an evolution of a Fageol design, but the new company soon established its own identity. By the mid-1950s, when its 281/351 series appeared, Peterbilt was one of the foremost US truck manufacturers.

**FRONT VIEW**

**REAR VIEW**

**Twin-wheeled** rear axle

**Fifth-wheel** trailer coupling

**Rubber suppression** between trailer and mudguard

**Trailing** rubber mud flaps

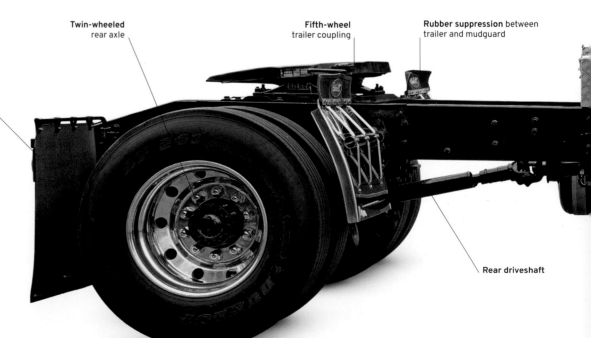

**Peterbilt 281**
This tractor unit was converted into a 6x4 during its working life, and used as a shot-blasting rig – a truck fitted with a high pressure compressor and blasting equipment to remove rust from metal surfaces. In 1999, a new owner restored the truck, reinstating the characteristic needle nose.

**Rear driveshaft**

THE 281 GAINED THE NICKNAME "Needle Nose" because of its narrow front and tapered hood. Large numbers of 281s were bought, both by fleet operators and owner/drivers. This model was the last Peterbilt truck to feature side-opening panels for the hood, before a one-piece, front-lifting hood became standard, offering greatly improved access to the engine area. Originally, the 281 was built as a 4×2 configuration (four wheels and one driven axle), while its sister truck, the 351, was a 6×4 (six wheels and two driven axles), with twin rear axles. The Peterbilt 358 was launched in 1965 to replace the 281/351 series, but the previous models continued to be manufactured until 1976. The axles, engine, and transmission all varied during production of the 281/351 series, but the classic design remained a constant throughout the entire run. A 281/351 model won cinematic fame as the "villain" in the 1971 movie *Duel*.

| SPECIFICATIONS | |
| --- | --- |
| Model | 281 |
| Origin | USA |
| Assembly | Oakland, California |
| Production run | 1954–1976 |
| Weight | 5.4 tonnes (5.4 tons) |
| Load capacity | Unknown |
| Engine | 6-cylinder turbo-diesel, 262 hp |
| Transmission | 10-speed manual |
| Maximum speed | 116 km/h (72 mph) |

**All-new brand identity**
The 281 was the first Peterbilt to use the famous red oval as the brand signature. Since 1953, the logo has become synonymous with these iconic American trucks.

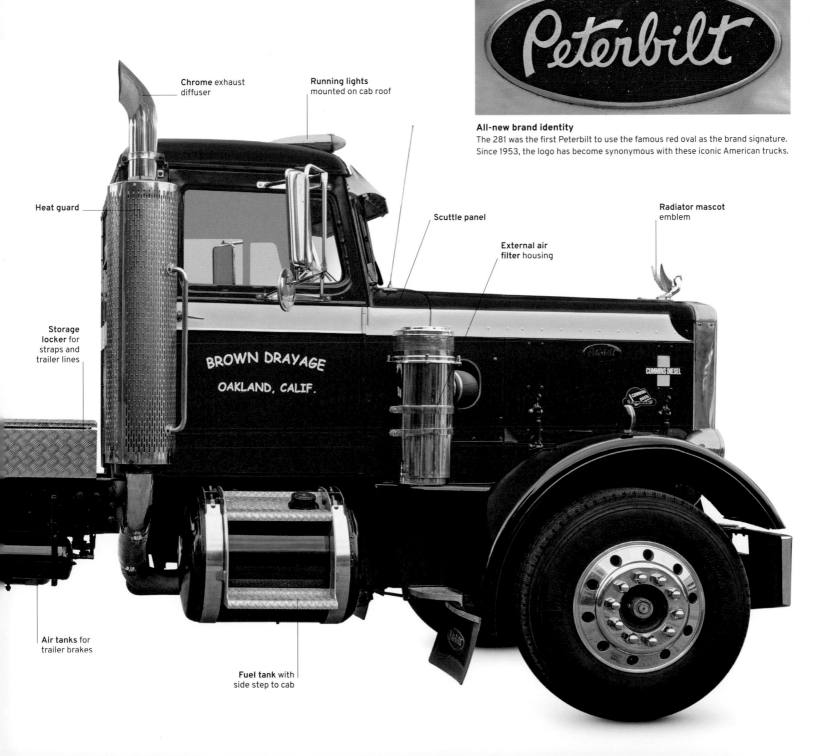

**Chrome** exhaust diffuser

**Running lights** mounted on cab roof

**Heat guard**

**Storage locker** for straps and trailer lines

BROWN DRAYAGE
OAKLAND, CALIF.

**Scuttle panel**

**External air filter** housing

**Radiator mascot** emblem

**Air tanks** for trailer brakes

**Fuel tank** with side step to cab

## THE EXTERIOR

This Peterbilt 281 has been fully restored to its original specifications, and painted in the red-and-white livery of its first operator, Brown Drayage of Oakland, California. Its cab, hubs, and wheels are made of aluminium to reduce the weight of the vehicle. The rear axle is a Rockwell SQHD. Rockwell are renowned for producing heavy-duty driving axles for trucks.

**1.** Radiator mascot  **2.** Cab corner step  **3.** 24-volt headlamp  **4.** Fuel tank with cab step mounted over top  **5.** Door handle to cab  **6.** Air tank for truck braking system  **7.** Aluminium hubs (with left-handed thread)  **8.** Sliding rear window in cab

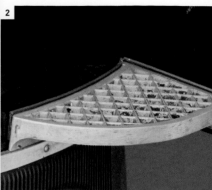

## THE ENGINE

Originally this Peterbilt left the Oakland factory with a Cummins HRB normal-aspirated diesel engine, which produced 165 hp. Later in the truck's life, and with advancements in engine power, a Cummins NTO turbodiesel engine was installed, which had an output of 262 hp. This engine pulls very well in the lightweight 281 series. The 10-speed RT910 transmission is made by Fuller.

**9.** Engine turbocharger  **10.** Engine bay showing the Cummings NTO 262 hp turbo-diesel engine

## THE INTERIOR

The cab's wooden floor can be easily removed to inspect the drivetrain and other under-cab components. The dashboard sits immediately ahead of the steering wheel, with a panel of switches for lights, ignition, and fuel supply to the right. The foot pedals are offset towards the driver. An engine decompression lever primes the engine for starting. The transmission lever has a selector switch at the top.

**11.** Cab dashboard and driving controls  **12.** Interior lamp in cab roof  **13.** Engine cut-off controls  **14.** Indicator lamp selector  **15.** Parking brake lever  **16.** Accelerator and brake pedals  **17.** Transmission lever

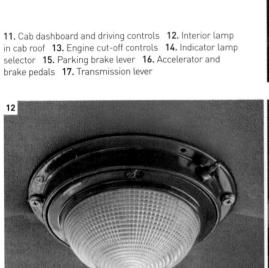

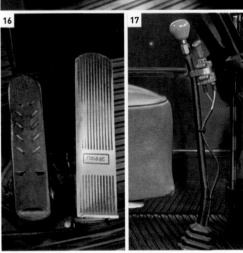

Dart oil truck, 1940s

# Key Manufacturers
# The PACCAR Story

The PACCAR story began in 1905 in Washington, where William Pigott, Sr manufactured rail and logging equipment. He could never have envisaged how big the PACCAR name was to become in the trucking world. Today, Kenworth, Peterbilt, and DAF trucks are all under the umbrella of this global, technology-leading brand.

**WILLIAM PIGOTT, SR** was born to Irish immigrant parents in Syracuse, New York, in 1860. As a child, his family moved to Ohio where the steel business was a dominant industry, and, he ended up working for the local mill as a salesperson.

Pigott moved to the Pacific Northwest in the 1890s to set out in business himself. The logging and railroad industries were booming there, and Pigott saw the opportunity to use his knowledge of the steel industry to supply them.

Pigott established the Railway Steel and Supply Company in 1901 to build horse-drawn logging trucks, structural steel for building works, and rails. In 1905, this venture was folded into his newly formed Seattle Car Company, which manufactured heavy-duty boxcars for logging railroads.

In 1917, The Seattle Car Company merged with the rival Twohy Brothers Company, creating the Pacific Car and Foundry Company, a name that stuck for 55 years. Pigott stepped down as president when the business was acquired by the American Car and Foundry in 1924, but stayed active in community affairs until his death.

In 1934, five years after Pigott's death, his son, Paul Pigott, became a major investor and company

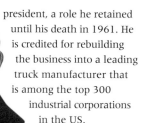

**William Pigott, Sr,**
(1860–1929)

president, a role he retained until his death in 1961. He is credited for rebuilding the business into a leading truck manufacturer that is among the top 300 industrial corporations in the US.

Under Paul's leadership, Pacific Car purchased the Kenworth Motor Truck Company in 1945, and both Dart Motor Trucks and Peterbilt Motors Company in 1958. Kenworth specialized in custom-built trucks, especially heavy-duty off-road models, designed for everything from off-highway logging operations to haulage over treacherous mountain and desert terrains. Dart made huge off-road trucks that brought Pacific Car into the mining industry. Peterbilt was known for tractor trucks – their

popular Model 351, nicknamed the "Needle Nose" for its narrow hood, launched in 1954, and remained in production for two decades.

Pacific Car became an international company in 1960 after moving into Mexico through a co-venture with truck manufacturer VILPAC, S.A. Its manufacturing operations expanded into Australia in 1966 with a new Kenworth factory. Kenworth's iconic

conventional-cab W900, a highway-going semi, had launched five years earlier in 1961 and is still available today – one of the longest production runs in truck history.

The need to provide belts, hoses, adapters, and other parts for the Kenworth and Peterbilt brands led to the creation of Dynacraft in 1967. The division still supplies parts and accessories, such as valves, battery boxes, emission systems, and dash panels that are produced in three facilities located in Washington, Kentucky, and Texas.

Pacific Car was officially renamed PACCAR in 1972, with PACCAR Parts forming in 1973 to provide sell and distribute parts for medium- and heavy-duty trucks, trailers, buses, and engines. They currently operate a network of 18 global parts distribution centres, selling in 115 countries.

In 1976, Kenworth set a benchmark for driver comfort with the debut of its Aerodyne sleeper cab, with a raised roof that offered more than 2 m (7 ft) of headroom. The company innovated again with the launch of a more

In 2022, **Kenworth** and **Peterbilt** achieved **six Manufacturing Leadership Awards** from the National Association of Manufacturers.

**Early Kenworth truck assembly**
Workers in a 1920s manufacturing facility assemble Kenworth 145A trucks. Designed to haul and dump gravel, the 145A had a 5.9-tonne (5.8-ton) gross vehicle weight.

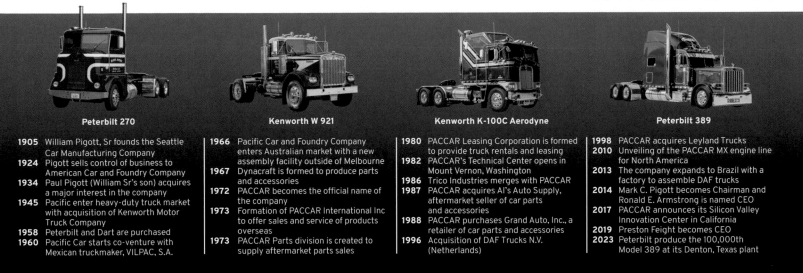

**Peterbilt 270**

| | |
|---|---|
| **1905** | William Pigott, Sr founds the Seattle Car Manufacturing Company |
| **1924** | Pigott sells control of business to American Car and Foundry Company |
| **1934** | Paul Pigott (William Sr's son) acquires a major interest in the company |
| **1945** | Pacific enter heavy-duty truck market with acquisition of Kenworth Motor Truck Company |
| **1958** | Peterbilt and Dart are purchased |
| **1960** | Pacific Car starts co-venture with Mexican truckmaker, VILPAC, S.A. |

**Kenworth W 921**

| | |
|---|---|
| **1966** | Pacific Car and Foundry Company enters Australian market with a new assembly facility outside of Melbourne |
| **1967** | Dynacraft is formed to produce parts and accessories |
| **1972** | PACCAR becomes the official name of the company |
| **1973** | Formation of PACCAR International Inc to offer sales and service of products overseas |
| **1973** | PACCAR Parts division is created to supply aftermarket parts sales |

**Kenworth K-100C Aerodyne**

| | |
|---|---|
| **1980** | PACCAR Leasing Corporation is formed to provide truck rentals and leasing |
| **1982** | PACCAR's Technical Center opens in Mount Vernon, Washington |
| **1986** | Trico Industries merges with PACCAR |
| **1987** | PACCAR acquires AI's Auto Supply, aftermarket seller of car parts and accessories |
| **1988** | PACCAR purchases Grand Auto, Inc., a retailer of car parts and accessories |
| **1996** | Acquisition of DAF Trucks N.V. (Netherlands) |

**Peterbilt 389**

| | |
|---|---|
| **1998** | PACCAR acquires Leyland Trucks |
| **2010** | Unveiling of the PACCAR MX engine line for North America |
| **2013** | The company expands to Brazil with a factory to assemble DAF trucks |
| **2014** | Mark C. Pigott becomes Chairman and Ronald E. Armstrong is named CEO |
| **2017** | PACCAR announces its Silicon Valley Innovation Center in California |
| **2019** | Preston Feight becomes CEO |
| **2023** | Peterbilt produce the 100,000th Model 389 at its Denton, Texas plant |

aerodynamic cab in 1985, with the T600A, improving fuel efficiency by 20 per cent over similar models.

PACCAR became a leader in the Mexican truck market with the purchase of VILPAC, S.A. in 1995. In Europe, PACCAR acquired DAF Trucks in the Netherlands in 1996 and Leyland Trucks in the UK in 1998. Expansion into international markets continued in 2013 when PACCAR opened a new factory in Ponta Grossa, Brazil.

In 2010, PACCAR invested more than $400 million in their engine factory and technology centre in Mississippi, launching its 12.9-litre, six-cylinder MX engine line that same year. PACCAR moved into the tech hub of Silicon Valley in 2017 to launch a new Innovation Center tasked with advancing the safety, sustainability,

**The Kenworth 853**
Kenworth designed this purpose-built truck in 1947 for a massive desert construction project, the Trans-Arabian oil pipeline.

and performance of heavy-duty trucks. The facility is a hub for the development of technologies, such as alternative fuels, autonomous driving systems, and the use of data analytics.

PACCAR's huge catalogue of trucks features six DAF ranges that have taken the International Truck of the Year title – including back-to-back wins in 2022 and 2023 for their ultra-efficient XF, XG, and XD Series designed for long haulage. The sister manufacturers

Kenworth and Peterbilt continue to have loyal followings in the trucking community. Kenworth introduced the next generation of its Class-8 flagship truck, the T680, in 2021, while Peterbilt announced the production of its 100,000th Model 389 in 2023. After more than a

century of success selling products in more than 100 countries, PACCAR continues to be a global leader in truck manufacture. Today, the company is partnering with other top companies, such as Toyota and Daimler Trucks, to accelerate battery and hydrogen fuel cell production.

**Stacked Peterbilt trucks**
A Peterbilt 365 Class-8 truck hauls identical cabs. Launched in 2007, the 365 retains classic styling.

# US Pickups that Made Detroit Hum

From the early days of the Ford Model T, passenger cars were produced in the Michigan city of Detroit in their millions by the "Big Three": Ford, Chrysler, and General Motors. Pickup trucks were added to their lines, and, from the late 1930s into the 1950s, as these smaller trucks became highly popular, large, established truck manufacturers also tried their hand at offering a smaller truck. However, they struggled to match production output at the same low costs the "Big Three" were able to achieve.

**Cab design** is carried over into wartime production

**Split screen** windscreen

### △ Mack ED

**Date** 1939   **Origin** USA

**Engine** Straight six, 67 hp

**Payload** 750 kg (1,650 lb)

Known for building heavy industrial trucks, Mack's ED was an attempt to tap into the light truck market. It struggled to take on its competitors despite being more than capable.

### △ Plymouth PT-125

**Date** 1941   **Origin** USA

**Engine** Straight six, 87 hp

**Payload** 500 kg (1,100 lb)

The PT-125 shared its front with a Plymouth saloon, but it was also based on a Dodge – both were part of the Chrysler group. This was a marketing ploy: to sell a Dodge, with a posh badge.

### ▽ Ford F-150

**Date** 1948   **Origin** USA

**Engine** Straight six, 95 hp

**Payload** 500 kg (1,100 lb)

The birth of a legend, the new F-150 was Ford's first completely new truck since World War II. It was also available as a panel van, which are now extremely rare.

**16 x 4.5 inch** steel wheels

### ▽ Studebaker M5

**Date** 1948   **Origin** USA

**Engine** Straight six, 80 hp

**Payload** 750 kg (1,650 lb)

Studebaker, known for its spacious station wagons, entered the pickup market in 1937 and in 1941 launched the M5 to rival Dodge, Ford, and Chevrolet. It was built to a robust budget and in the postwar truck boom, it sold well at $1,082.

**Optional extra** chrome grille guard

Headlamps powered by 12-volt electrics

235/85R/16 tyres

◁ **Diamond T 201**

**Date** 1949  **Origin** USA

**Engine** Straight six, 91 hp

**Payload** 3.6 tonnes (4 tons)

The instantly recognizable design of a Diamond T has a sense of Art Deco flare. Ohio company Hercules provided the engine power, a partnership that served the brand well.

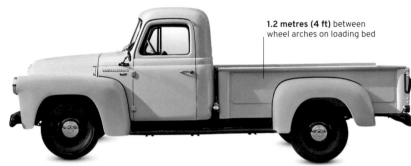

1.2 metres (4 ft) between wheel arches on loading bed

▽ **GMC Half Ton Long Bed**

**Date** 1953  **Origin** USA

**Engine** Straight six, 95 hp

**Payload** 500 kg (1,100 lb)

The postwar GMC pickup reflected a more aerodynamic approach than the previous late-1930s cabs. For many returning servicemen, a GMC would be the only truck they would buy.

▷ **International S-112**

**Date** 1956  **Origin** USA

**Engine** 80 hp

**Payload** 500 kg (1,100 lb)

A pioneer among America's great truck manufacturers, International produced a range of pickups in the 1950s. The Ohio firm marketed the S-series light truck as having the "roomiest, most comfortable cab on the road".

▽ **Ford F-250 4x4**

**Date** 1959  **Origin** USA

**Engine** V8, 172 hp

**Payload** 2.5 tonnes (2.7 tons)

As a 4x4 with a greater payload, the F-250 appealed to wilderness drivers, particularly in Canada. Ford had previous experience with all-wheel drive during large-scale military truck production in World War II.

Pressed rear panels incorporating 50s fin styling

△ **Chevrolet El Camino**

**Date** 1959  **Origin** USA

**Engine** V8, 246–330 hp

**Payload** 290–520 kg (640–1,150 lb)

A cross between a fin-tailed cruiser and a utility vehicle, the El Camino proved to be a great success because of its unique style and the station wagon platform it was based on. It was one of the fastest pickups of the era.

**Later** after-market chrome wheels

# Timber Trucks Hit the Road

In the 1930s, as road building increased, the development of heavy-duty trucks took off, too. As diesel-powered trucks grew to be standard, they became cheaper to run. The larger, more powerful engines of this period also allowed for hauling heavier loads at higher speeds over greater distances. Brakes were improved and solid rubber tyres were replaced by inflated pneumatic tyres with treads that could handle rougher roads. As trucks could now reach previously inaccessible areas, they were an obvious choice for logging companies, which had been heavily reliant on rail in the past. By 1931, around 900 trucks were operating in Oregon and Washington, the centre of the US timber industry, and by the 1950s, logs were hauled almost exclusively by increasingly powerful trucks.

## TALKING THE TORQUE

The 1950s brought a new generation of heavy-duty trucks to the highways and logging roads. With them came the shift to turbocharged diesel engines, giving trucks a much greater power-to-weight ratio. Gross vehicle weights, which combine a truck and its load, on off-highway models now rose as high as 45 tonnes (44 tons). The newly fitted diesel engines also generated higher torque, which is the real indicator of towing power, making them ideal for timber operations.

**A 1950s Mack truck from the off-highway range of the L-series** digs deep as it transports an enormous load of timber in the Pacific Northwest.

# Trucks that Rebuilt Britain

During the 1950s, the UK could boast a great number of lorry makers. Scammell, Albion, AEC, and Thornycroft were all absorbed into the Leyland brand (later becoming Leyland DAF). Others, such as Vulcan, Commer, Seddon Atkinson, Bedford, and Foden have now vanished. British manufacturers played a key role in the development of the lorry, pioneering innovations such as cab-over and air suspension. Foden was the first firm in the world to produce a tilt-cab, a feature that is now virtually universal on European heavy-goods vehicles.

**Hydraulic** tipping carry tray

### △ Bedford O Type

**Date** 1946  **Origin** UK

**Engine** Inline 4- or 6-cylinder petrol, 72 hp

**Payload** 5–6 tonnes (5–6 tons)

The US's General Motors established a plant in Luton, UK in 1931. With its distinctive bull nose, the 5-ton O-series had a four-speed gearbox and a spiral bevel rear axle for a smoother drivetrain.

**Headboard**

### ▽ Thornycroft 3-ton Nippy Truck

**Date** 1949  **Origin** UK

**Engine** Inline 4-cylinder, 60 hp

**Payload** 3 tonnes (3 tons)

Known as "the Nippy" because of its manoeuvrability, this truck was ideal for operating in towns and cities. Available as a flat bed or a box van, it delivered everything from bags of coal to sacks of fruit.

**Side mirror** attached to the a-pillar

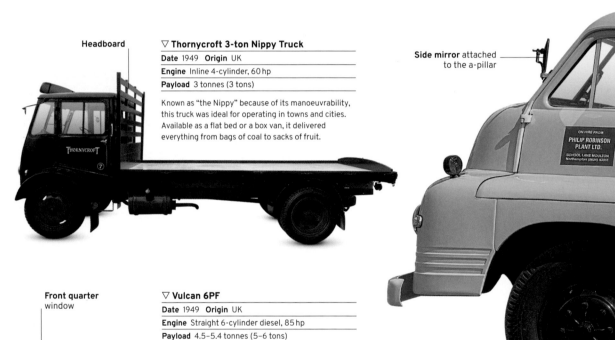

ON HIRE FROM
PHILIP ROBINSON
PLANT LTD.
SCHOOL LANE MOULTON
Northampton (0604) 43863

**Front quarter** window

### ▽ Vulcan 6PF

**Date** 1949  **Origin** UK

**Engine** Straight 6-cylinder diesel, 85 hp

**Payload** 4.5–5.4 tonnes (5–6 tons)

The Vulcan 6PF was a premium middleweight lorry equipped with a Perkins diesel engine that was famously difficult to start in cold weather. The 6VF was petrol powered.

**Canvas** canopy

**Fuel** tank

### ▷ Bedford RL

**Date** 1953  **Origin** UK

**Engine** Straight 6-cylinder petrol, 110 hp

**Payload** 3–4 tonnes (3–4 tons)

The Bedford RL was the workhorse of the British Army from the 1950s to the 1970s. Based on the civilian Bedford SCL, it could be equipped with large diameter wheels and four-wheel drive for improved performance over rough ground.

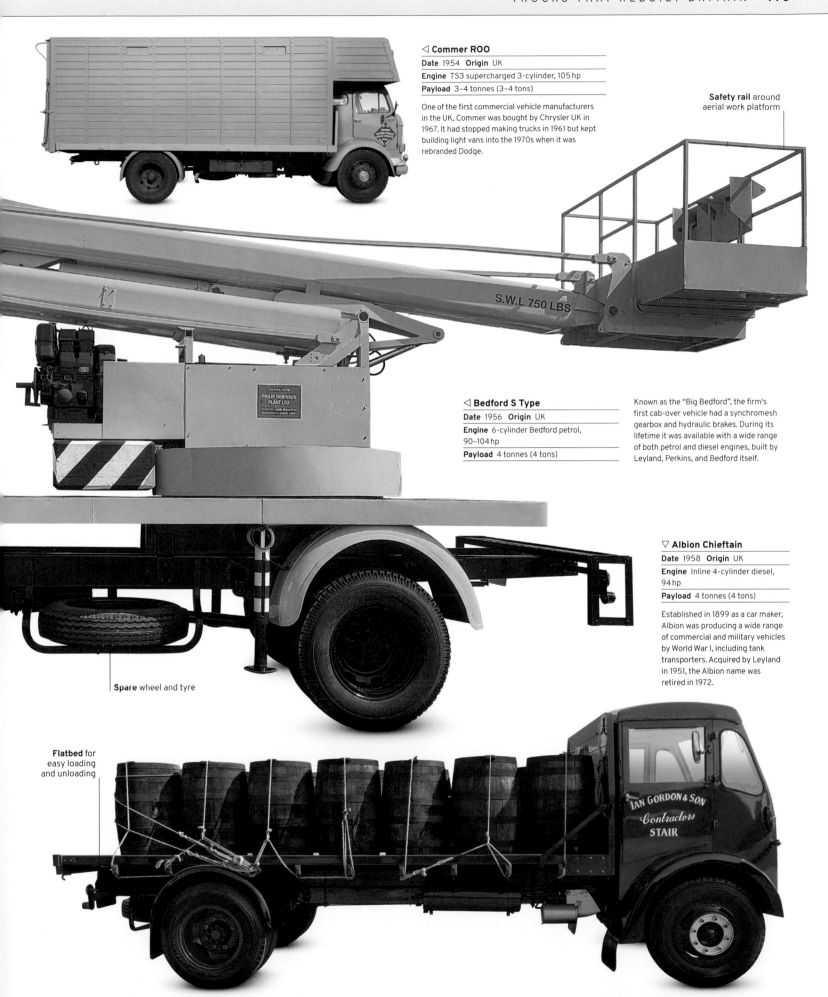

◁ **Commer ROO**

**Date** 1954 **Origin** UK

**Engine** TS3 supercharged 3-cylinder, 105 hp

**Payload** 3–4 tonnes (3–4 tons)

One of the first commercial vehicle manufacturers in the UK, Commer was bought by Chrysler UK in 1967. It had stopped making trucks in 1961 but kept building light vans into the 1970s when it was rebranded Dodge.

**Safety rail** around aerial work platform

S.W.L 750 LBS

◁ **Bedford S Type**

**Date** 1956 **Origin** UK

**Engine** 6-cylinder Bedford petrol, 90–104 hp

**Payload** 4 tonnes (4 tons)

Known as the "Big Bedford", the firm's first cab-over vehicle had a synchromesh gearbox and hydraulic brakes. During its lifetime it was available with a wide range of both petrol and diesel engines, built by Leyland, Perkins, and Bedford itself.

**Spare** wheel and tyre

▽ **Albion Chieftain**

**Date** 1958 **Origin** UK

**Engine** Inline 4-cylinder diesel, 94 hp

**Payload** 4 tonnes (4 tons)

Established in 1899 as a car maker, Albion was producing a wide range of commercial and military vehicles by World War I, including tank transporters. Acquired by Leyland in 1951, the Albion name was retired in 1972.

**Flatbed** for easy loading and unloading

1960s
# THE GOLDEN AGE

## THE EXTERIOR

During the late 1960s, a forward-opening fibreglass hood was marketed to improve accessibility to the engine compartment; in doing so, the design of this B61 was changed from the standard appearance to incorporate a larger, more mainstream radiator grille and twin headlamps. Another differentiating feature of this truck is the large front bumper, which was not standard on the typical B61 series truck. A vertical exhaust stack extends up the right-hand side of the driver's cab.

**1.** Bulldog symbol of Mack brand  **2.** Thermodyne engine badge  **3.** Wheel  **4.** Hood clamp  **5.** Grab handle to cab  **6.** Bulldog radiator/hood mascot  **7.** Indicator  **8.** Running light mounted on cab roof  **9.** Air tanks for brakes  **10.** Suspension airbag

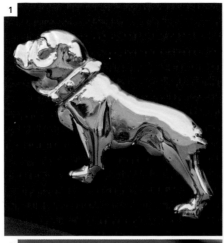

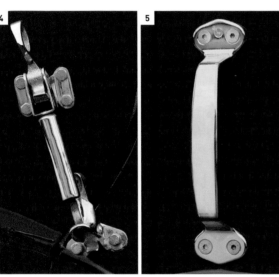

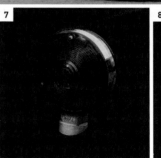

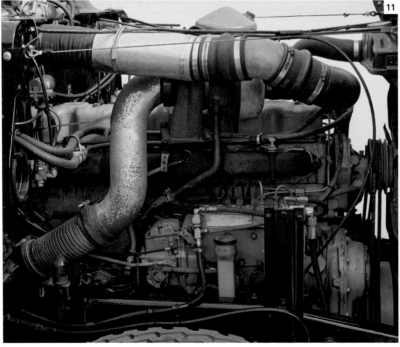

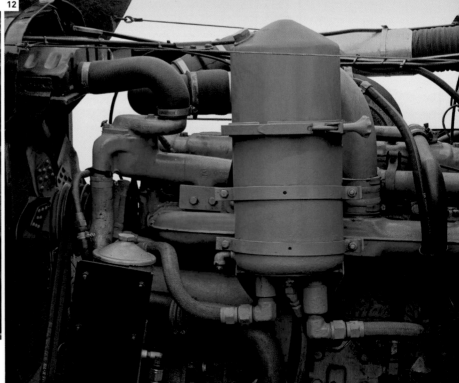

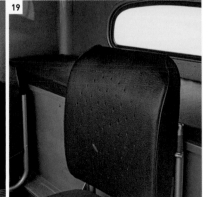

## THE ENGINE

Mack's new diesel engine – the open chamber, direct-injection Thermodync – was a revolution when it was launched in the B series in 1953. A brand-new engine in a brand-new vehicle always risks unwanted teething troubles, but it proved a success. The engine shown here is a 6-cylinder, turbocharged, 11-litre Thermodyne. Producing 220 hp at between 1,500–2,000 revs, the engine was less powerful than the units in many rival trucks. However, through timely gear changes, the driver could make steady progress on most road surfaces when hauling a load.

11. Engine, showing turbocharger (top centre) and diesel-injector pump (bottom centre)
12. Oil filter

## THE INTERIOR

Mack trucks were built for work, and the spartan furnishings of the B61's cab reflect this. Eye-catching green paintwork enhances the stylish-yet-functional cab design. The floor is covered with rubber matting, which is easy to remove and clean. The clever dashboard layout allows for a simple conversion from left-hand drive to right by swapping the positions of the instrument panel and glove compartment.

There are two transmission levers, low- and high-range, both for on-road use. This being a sleeper cab, a rather rudimentary single bed is provided behind the seats.

13. View of cab interior  14. Air-mixture control for engine
15. Cooling and suspension controls  16. Indicator selector switch  17. Passenger seat  18. Cast-aluminium throttle pedal  19. Rear bed  20. Cab door interior panel

# The Changing Face of Pickups

While still continuing to push the boundaries with innovative, forward-thinking designs, the motor manufacturers of the 1960s also offered pickups based on existing cars. Many smaller, historic brands had been swallowed up by the big corporate giants, which led to cost-sharing between models, and separate chassis cabs were fast becoming a thing of the past (except for purpose-built, heavy-duty pickups). The new car-based pickups proved ideal vehicles for light work, with the bonus of passenger comforts and the ability to fit into conventional parking spaces.

### △ Holden FB

**Date** 1961  **Origin** Australia

**Engine** 6-cylinder, 75 hp

**Payload** 1 tonne (1 ton)

The popular B series included saloon, station wagon, panel van, and pickup. Holden is owned by General Motors; design and parts-sharing between models explain the FB's Chevy look.

**Forward-opening** door with hinged steering column

**Sun roof** also serves as an emergency exit

### △ BMW Isetta

**Date** 1961  **Origin** Germany/UK

**Engine** 1-cylinder, 37 hp

**Payload** 74 kg (164 lb)

Based on the Isetta microcar, this tiny pickup was used by the UK's RAC breakdown service for roadside repairs. Available in 3-wheel or 4-wheel versions, it used a BMW motorcycle engine.

### △ DAF 750 Pickup

**Date** 1961  **Origin** Netherlands

**Engine** 2-cylinder, air-cooled, 26 hp

**Payload** 330 kg (724 lb)

Based on the DAF 33 saloon, the DAF 750 was used for light deliveries and proved a popular work vehicle for engineers. It is notable for pioneering a continuously variable transmission.

**Drop-down** tailgate

**Chrome bumpers**

**36-cm (14-inch)** pressed steel wheels

### △ Chevrolet Corvair 95 Rampside

**Date** 1962  **Origin** USA

**Engine** Air-cooled, flat-six cylinder, 80 hp

**Payload** 750 kg (1,653 lb)

Taking up no more space than a station wagon, this truck was intended to rival the pickup version of VW's Type 2 van. As well as a tailgate, it had a gate on its right side that could double up as a ramp.

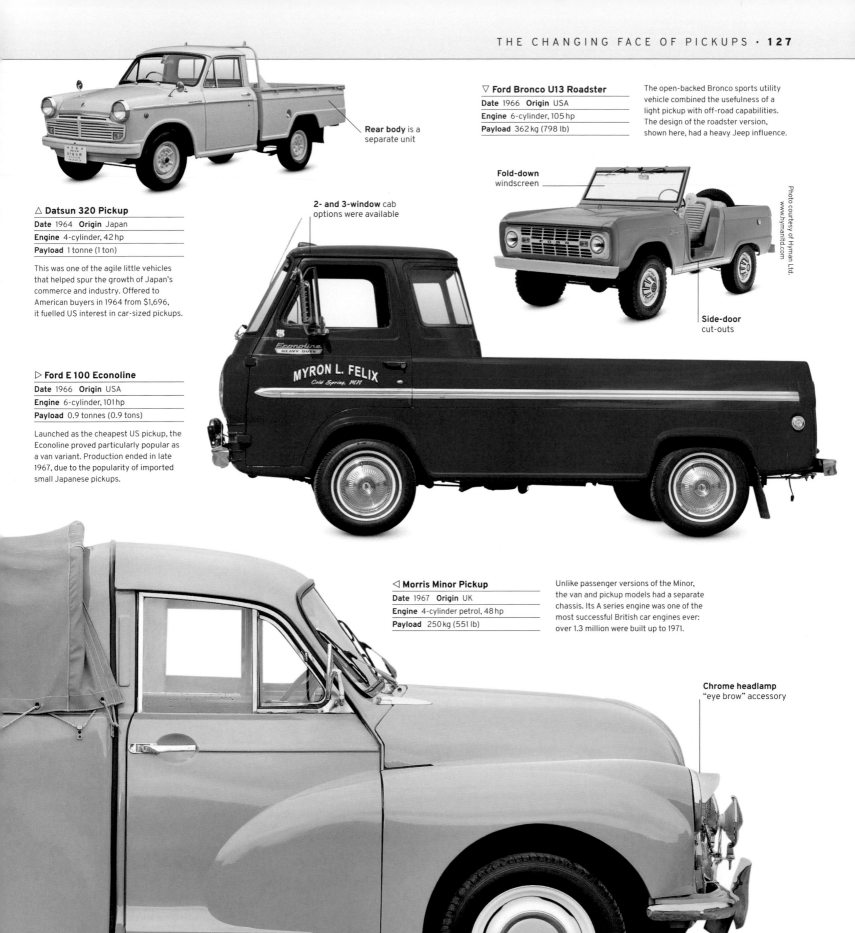

Rear body is a
separate unit

### △ Datsun 320 Pickup

**Date** 1964 **Origin** Japan

**Engine** 4-cylinder, 42 hp

**Payload** 1 tonne (1 ton)

This was one of the agile little vehicles
that helped spur the growth of Japan's
commerce and industry. Offered to
American buyers in 1964 from $1,696,
it fuelled US interest in car-sized pickups.

### ▷ Ford E 100 Econoline

**Date** 1966 **Origin** USA

**Engine** 6-cylinder, 101 hp

**Payload** 0.9 tonnes (0.9 tons)

Launched as the cheapest US pickup, the
Econoline proved particularly popular as
a van variant. Production ended in late
1967, due to the popularity of imported
small Japanese pickups.

### ▽ Ford Bronco U13 Roadster

**Date** 1966 **Origin** USA

**Engine** 6-cylinder, 105 hp

**Payload** 362 kg (798 lb)

The open-backed Bronco sports utility
vehicle combined the usefulness of a
light pickup with off-road capabilities.
The design of the roadster version,
shown here, had a heavy Jeep influence.

Fold-down
windscreen

Photo courtesy of Hyman Ltd.
www.hymanltd.com

Side-door
cut-outs

2- and 3-window cab
options were available

### ◁ Morris Minor Pickup

**Date** 1967 **Origin** UK

**Engine** 4-cylinder petrol, 48 hp

**Payload** 250 kg (551 lb)

Unlike passenger versions of the Minor,
the van and pickup models had a separate
chassis. Its A series engine was one of the
most successful British car engines ever:
over 1.3 million were built up to 1971.

Chrome headlamp
"eye brow" accessory

1932 Dodge half-ton panel truck
launched soon after the company's
acquisition by Chrysler

# Key Manufacturers
# The Dodge Story

The Dodge brand originated in the US in the early 1900s with two brothers. They were among the early automobile manufacturers competing in the new era of transportation, as horse-drawn carts gave way to petrol and diesel-powered freight. Dodge's legacy of powerful, oversize pickups is carried on today by Ram Trucks.

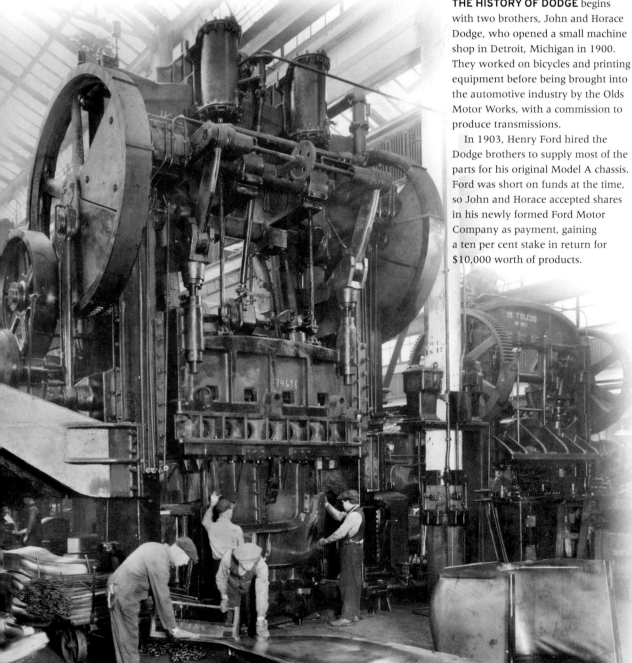

**THE HISTORY OF DODGE** begins with two brothers, John and Horace Dodge, who opened a small machine shop in Detroit, Michigan in 1900. They worked on bicycles and printing equipment before being brought into the automotive industry by the Olds Motor Works, with a commission to produce transmissions.

In 1903, Henry Ford hired the Dodge brothers to supply most of the parts for his original Model A chassis. Ford was short on funds at the time, so John and Horace accepted shares in his newly formed Ford Motor Company as payment, gaining a ten per cent stake in return for $10,000 worth of products.

**Dodge LC "Humpback" Panel Truck**
This popular half-ton delivery truck showed off the sleek curves and contours of Dodge's new line of commercial trucks launched in 1936.

In 1914, the Dodge brothers used their Ford profits to break away and introduce their own first car, the Model 30–35, under the newly established Dodge Brothers Motor Company. Their four-cylinder touring car earned a reputation for quality and durability, and by 1916, Dodge ranked second in sales in the US.

In 1918, Dodge heeded the call from its sales team to bring out its first commercial truck. Called a "Dodge Brothers Business Car", it was a canopy-covered delivery truck with screens on the sides and a 450 kg (1,000 lb) payload.

After both Dodge brothers passed away in 1920 during a flu epidemic, their widows put Frederick Haynes in charge of the company until it was sold to an investment group for $146 million in 1925. The Chrysler Corporation then acquired Dodge for

**Press room at Dodge Main, 1918**
The Dodge Main factory opened on a 24-acre site in Hamtramck, Michigan, in 1910. Here, workers are producing pieces of the pressed steel bodies that replaced wooden frames.

**Dodge Brothers Canopy Truck**

**Dodge Power Wagon**

**Dodge Rampage**

**Dodge SRT**

| | | | | | | | |
|---|---|---|---|---|---|---|---|
| **1900** | Dodge Brothers Company is formed by Horace and John Dodge in Detroit, MI | **1926** | Graham Brothers, a body manufacturer, is fully acquired by Dodge | **1960** | The Dodge LCF series of medium- and heavy-duty trucks launch | **1998** | Dodge is first to offer 4-door pickup |

**1900** Dodge Brothers Company is formed by Horace and John Dodge in Detroit, MI
**1903** Henry Ford commissions parts from the Dodge brothers, who gain a 10 per cent stake in his company as payment
**1910** The Dodge Main manufacturing facility opens
**1914** Dodge sells its first car
**1920** The Dodge brothers both die after catching flu at an auto exhibition
**1925** The Dodge Brothers Motor Company is sold to an investment group

**1926** Graham Brothers, a body manufacturer, is fully acquired by Dodge
**1928** Dodge Brothers Motor Company is sold to Chrysler Corporation
**1936** Dodge enters large truck market with the MD series
**1939** Dodge begins "job-rating" its heavier trucks
**1940** The WC/WD series of four-wheel drive military truck is introduced
**1946** Dodge manufactures its first heavy-duty diesel engine truck

**1960** The Dodge LCF series of medium- and heavy-duty trucks launch
**1964** The Dodge Van is introduced
**1976** Dodge halts production of heavy-duty trucks
**1974** The Dodge Ramcharger SUV debuts, originally with four-wheel drive
**1979** Chrysler is saved from bankruptcy with a government loan
**1981** D series rebranded as Dodge Ram
**1987** Chrysler acquires American Motors and assumes the Jeep® brand

**1998** Dodge is first to offer 4-door pickup
**1998** Chrysler is merged with Daimler-Benz AG and renamed DaimlerChrysler
**2004** The SRT-10 pickup is introduced with a Viper engine
**2009** The company is reorganized as Chrysler Group LLC, and Dodge's trucks division is rebranded as Ram
**2014** Chrysler Group LLC merges with Fiat SpA to become the FCA Group
**2023** The Ram 1500 REV electric concept truck is introduced

**Dodge 2-Ton Stake advertisement, 1941**
From 1939–41, Dodge introduced their T, V, and W series of "Job-Rated" trucks, such as the 2-Ton Stake. These came in a huge array of models, configured for specific types of work.

$170 million in 1928 in a move to compete against Ford and Chevrolet in the low-priced vehicle market.

The last Dodge-Brothers-designed truck, called the Merchant Express, came out in 1929. It was a half-ton pickup with four-wheel hydraulic brakes and options for a 45-, 63-, or 78-hp engine. Dodge trucks were produced alongside Chrysler's DeSoto and Plymouth labels at the time, helping Chrysler overtake Ford to be second in sales by 1936.

Until 1936, Dodge trucks were built on car chassis, but this changed with the development of heavier-duty trucks. In 1939, Dodge began marketing "job-rated" trucks, meaning that each was engineered specifically for the type of haulage being performed. The huge number of variants included options for up to 175 different chassis types.

In the lead-up to WWII, Dodge developed the US Army's first standard medium four-wheel-drive trucks. The company's military WC-series gave rise to the Power Wagon, a 4x4 medium-duty truck that was released in 1946 and remained in production until 1980.

The 1950s saw Dodge debut its first HEMI® V8 engine, the "Red Ram", and a fully automatic transmission. It also undertook a complete redesign of its trucks, launching the Dart Sweptline Pickup in 1961. Its sleek yet roomy modern styling and wider, lower body proved popular, and the "Sweptline era" lasted more than a decade before being supplanted by the D series.

Following a series or recessions and petrol shortages in the 1970s, Dodge's sales steeply declined. In 1979, Chrysler's chairman, Lee Iacocca, asked for a federal loan to keep the company from bankruptcy. After closing plants and reducing wages, the loan was repaid in 1983.

In 1981, the company rebadged their updated D series of pickups, which had first launched in 1972, as the Dodge Ram. The name was inspired by the ram-shaped bonnet emblem, symbolizing toughness and surefootedness, that appeared on the company's vehicles starting in 1931.

In 1998 Chrysler merged with Germany's Daimler-Benz AG. The renamed company, DaimlerChrysler AG, filed for bankruptcy protection following the global financial crisis of 2007–08, and was bailed out by the US Treasury. In the ensuing company reorganization, Dodge's line of pickups was rebranded as Ram Trucks, which has risen as high as second in the US truck market. Fiat purchased Chrysler shares from the US Government in 2011, merging to become Fiat Chrysler Automobiles (FCA) in 2014. FCA joined the conglomerate Stellantis, based in the Netherlands, in 2021 to become the third largest automotive manufacturer in the world.

**Ram 1500 TRX, 2021**
With its oversize proportions and 702-hp Hellcat V8 engine, the 2021 Ram 1500 TRX has robust off-road capabilities.

"Buying **Dodge** was one of the **soundest acts** of my life. I say sincerely that **nothing we have done… compares** with that transaction."
WALTER P. CHRYSLER, FOUNDER OF CHRYSLER CORPORATION

# On the Streets

1968 was a year of civil unrest in France. In May, students took to the streets demanding greater freedom and an end to the Vietnam War. The spirit of protest quickly spread to industry. Renault was one of the first companies to see its staff walk out, initially at its Cléon factory on 15 May, followed by workers at its plants at Le Mans, Boulogne-Billancourt, and Flins.

## MASS WALK-OUTS

The Renault walk-outs were a catalyst, triggering workers in other sectors to follow suit. Within a week, over 10 million workers had downed tools and economic life in France ground to a halt. While the workers' demands were different to those

of students – better pay, working conditions, and pensions – the two groups coalesced around a general sense of dissatisfaction with life and government in France. As protests mounted, including counter-protests by those loyal to president Charles de Gaulle, parliament was dissolved and elections were called. Although de Gaulle's party won, he resigned the next year. After May 1968, workers in many industries, including those at Renault and other automotive companies, received better pay and conditions, and the trade unions grew in strength.

**Workers from Renault**, riding in their company's trucks, drive through the streets of Paris in 1968 to make their feelings of discontent known.

# Novelty Trucks

Almost as soon as trucks took to the roads, their owners saw the opportunity to use them as mobile advertisements. Often it is the artwork on the trucks that catches the eye, but sometimes the trucks themselves are heavily customized or modified into all sorts of shapes and sizes in order to draw attention.

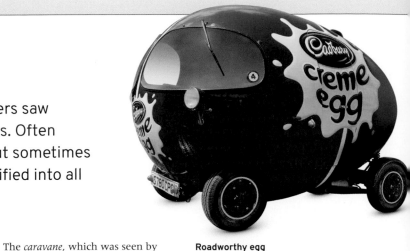

**Latil Tour de France van, 1935**
A lion, motif of the Argentil polish brand, sits on the roof of this van made by Latil. Striking artwork and lettering adorn the van's sides.

**Roadworthy egg**
Three Creme Egg vehicles toured the UK, promoting the soft-centred chocolate eggs of the Cadbury confectionery firm.

The first advertising on trucks was comparatively straightforward. The very earliest commercial vehicles had sign-written names, with the contact details, and sometimes the company's logo. In time, some companies also adopted corporate colours, which made their trucks easy to recognize on the roads. However, it was not long before truck advertising became far more creative. Adding a product-shaped body to a truck chassis proved an effective marketing tool.

In the first decades of the 20th century, Kodak created a camera-shaped truck that drove around the US with a "Kodak Girl" precariously balanced on top. Then, in the early 1920s, the Worthington brewery at Burton upon Trent, UK, bought five Daimler 30-hp trucks and converted them into bottle shapes. Bottle trucks were not just a British anomaly – in the US during the 1930s, a number of dairies made "bottle trucks" on a wide variety of chassis to deliver dairy goods.

## Caravan of love

What is better than the occasional novelty truck? A whole parade of them, of course. In 1929, the Tour de France started its *caravane publicitaire*, in order to help cover the expenses of staging the event. The idea was that officially endorsed advertising trucks would pay to accompany the cyclists. The parade soon became an essential part of the Tour de France.

By the end of the 1930s, the *caravane* had grown to a convoy of more than 60 colourful, noisy trucks promoting a wide range of brands through artwork, slogans, cutout signboards, and even giant figures attached to the roof. When the Tour resumed after World War II, the advertising trucks returned, bigger, brighter, and bolder.

The *caravane*, which was seen by millions each year as the Tour made its way through France, was a great way for companies to attract large numbers of potential customers. As a result, the procession grew ever larger and the vehicles more bold. Companies commissioned coach-makers to make unique, eye-catching vehicles in the shape of the products they were advertising, which led to everything from gas bottles to vacuum cleaners driving along the roads. In 1953, the Bic pen company caused a stir with a truck shaped like their iconic Cristal ballpoint pen.

> " The **Toe** is **my homerun**, became a **Seattle icon** and for me, that was **a big wow.**"
>
> ED LINCOLN, INVENTOR OF THE TOE TRUCK

The customizing tradition of the Tour de France *caravane* continues today, with the designs no less zany and attention-grabbing. While the cyclists may speed past, the *caravane* can now take well over 30 minutes to roll by.

## Assorted shapes

The idea of novelty trucks found lasting fame in the US in the shape of a sausage. The Wienermobile is a hot-dog-shaped truck that advertises Oscar Mayer frankfurters in the US. The first Wienermobile appeared in 1936, created by Oscar Mayer's nephew, Carl G. Mayer. Over the years, it has been through many incarnations, using different types of car, van, motorhome, pickup, and purpose-built chassis. Today, a fleet of these motorized sausages still cruise the highways – and they are so famous that toy Wienermobiles

and scale models have been made. Other creatively shaped trucks have included those that look like toothpaste tubes, pencils, shoes – even chocolate eggs. The Cadbury's Creme Egg truck was a UK oddity that first appeared in the late 1980s. Built on a Bedford Rascal van chassis, with a fibreglass body in the shape of an egg, it was painted to resemble the wrapper of the popular confectionery.

Among the most familiar of today's promotional vehicles are Coca-Cola's Christmas trucks. Although they feel older, the truck was first used in 1995. Illuminated and painted Santa-red, these trucks tour many countries in the run-up to Christmas.

**Do you want mustard on that?**
A Wienermobile visits the University of Oregon in Eugene, USA, in 2015. Drivers of Wienermobiles are known as "hotdoggers".

**TALKING POINT**

### Fun Pun Truck

In 1980, Ed Lincoln, owner of Lincoln Towing of Seattle in the US, built a tow truck with toes sprouting from the cab. He bought an old VW bus and commissioned a body shop to make a chicken wire and fibreglass cab. The toes (those of a left foot) were carved out of polystyrene foam. The promotional "toe truck" was painted bubblegum pink, and lights were installed in the toes to show up at night.

The truck was driven in parades and appeared at charity events. It could even be hired for weddings. In 1984, it was installed on the roof of the Lincoln Towing building, creating a landmark. A truck with the toes of a right foot was built in 1996.

**Lincoln's original "left" Toe Truck** was donated to Seattle's Museum of History and Industry. The "right foot" toe truck is still on the road.

# Vans for Every Use

One of the key advances in van design in the 1960s was the introduction of diesel engines for light commercial vehicles, giving them better fuel efficiency and increased power. Improved roads and easier international shipping in the 1960s made exporting new models more cost effective, opening up fresh sales opportunities across the globe. Among the most successful truck exports was Germany's Volkswagen Type 2 – its camper and pickup versions catching the imagination of American youth, eager to soak up everything that symbolized a life of freedom on the road.

Tarpaulin cover and frame can be removed for larger loads

Flat-faced cab mounted over the engine gives the driver a good view

Rear body made by Furgone coach builders of Madrid

Aerodynamic bonnet

### △ Iso Isettacargo

| | |
|---|---|
| Date 1960s | Origin Italy/Spain |

Engine 1-cylinder, two-stroke, 9.5 hp

Payload 500 kg (1,102 lb)

In 1954, Iso of Italy launched a van version of its Isetta microcar with a widened rear axle and extended wheelbase. Also built under licence in Spain, it had double side doors and a front-opening cab.

### △ DKW Schnellaster Kastenwagen

| | |
|---|---|
| Date 1960 | Origin Germany |

Engine 3-cylinder, two-stroke, 42 hp

Payload 750 kg (1,653 lb)

To maximize loading capacity, DKW's "Rapid Transporter" van housed the drivetrain at the front and stowed the spare wheel under the seat. Pickup and microbus versions were also available.

# Food on the Go

With the ever-growing number of public events, from baseball games to rocket launches and rock concerts, came the need for catering trucks to supply crowds with snacks and drinks. Meanwhile, companies continued making local deliveries, taking food to stores and homes, or, in the case of ice-cream makers, treats to the kerbside. Big trucks were too cumbersome for these tasks, but car-sized vans were ideal. Drivers could stop and serve wherever they saw custom, or make swift drop-offs, then move on to the next event or delivery address.

### ◁ Ford Ice Cream Truck

| | |
|---|---|
| Date 1965 | Origin USA |

Engine 6-cylinder petrol, 135 hp

Payload 520 kg (1,146 lb)

White Good Humor ice cream trucks were a happy summer sight in US neighbourhoods. There was only one door (complete with flashing safety sign) to this Ford, so the driver had to exit and serve on the kerbside.

△ **Mercedes-Benz L 319 Fire Tender**

| Date | 1961 | **Origin** Germany |
|------|------|------|
| **Engine** 4-cylinder diesel, 55 hp | | |
| **Payload** 2 tonnes (2 tons) | | |

In the 1960s, most fire appliances in Germany and the Netherlands were Mercedes L 319 fire tenders. The model shown here, however, was a race transporter, but has been restored and repurposed as a catering truck.

▽ **Volkswagen Type 2 Pickup**

| Date | 1962 | **Origin** Country |
|------|------|------|
| **Engine** 4-cylinder air-cooled, 46 hp | | |
| **Payload** 1 tonne (1 ton) | | |

The rear-engined Volkswagen pickup sold well across the world. This split-screen, crew-cab example has had its suspension lowered and Porsche fuch-style wheels added.

**Crew cab** has four doors and two rows of seats

**Drop-down** side panels

△ **Mercedes-Benz L 319 Pickup**

| Date | 1964 | **Origin** Germany |
|------|------|------|
| **Engine** 4-cylinder diesel, 55hp | | |
| **Payload** 2 tonnes (2 tons) | | |

This was the first Mercedes built postwar with global sales in mind. Aerodynamic lines and an upmarket interior won the L 319 a slice of the market for versatile, medium-sized commercial vehicles. Like this model, many pickups of the time had canvas-covered cargo beds.

**High roof line** enables standing inside for ease of loading

◁ **Mercedes-Benz L 407**

| Date | 1965 | **Origin** Germany |
|------|------|------|
| **Engine** 4-cylinder petrol, 80 hp | | |
| **Payload** 2 tonnes (2 tons) | | |

This "Racing Service" van carried spares and crew to race meetings. The skylights in the roof – a 1960s style feature – made the L 407 vans brighter inside. The L 407's last year of production was 1966.

Photo courtesy of Hyman Ltd. www.hymanltd.com

▷ **Divco Milk Truck**

| Date | 1965 | **Origin** USA |
|------|------|------|
| **Engine** 4-cylinder petrol, 38 hp | | |
| **Payload** 0.9 tonnes (0.9 tons) | | |

With a distinctive design dating back to 1937, Divco delivery vans featured a flat floor all the way through to allow swift loading and unloading. A stand-up driving position also enabled rapid exit and entry when making drop-offs.

△ **Renault Estafette Ice Cream Truck**

| Date | 1969 | **Origin** France |
|------|------|------|
| **Engine** 4-cylinder petrol, 60 hp | | |
| **Payload** 1 tonne (1 ton) | | |

This Estafette was Renault's answer to Citroën's H Van, which dominated the French market. A high roof allowed for standing inside. The engine, or a small generator, powered the chillers.

Photo courtesy of Hyman Ltd. www.hymanltd.com

# Citroën H Van

Thanks to its distinctive corrugated-steel exterior and "pig snout" front, the French Citroën H van became one of the most instantly recognizable small vans in mid-20th century Europe. Unveiled at the Paris motor show in 1947, the H caused a sensation, both for its looks and technical innovation. Crucial to the design was the monocoque (single shell) structure, with body and chassis integrated, making the van both light and strong.

**IN PRODUCTION FOR OVER 30 YEARS,** the Citroën H van offered various wheelbase options and purpose-made variants, which ranged from pickups to high-roof vans, and even ambulances. It was as a boxy, compact delivery van, however, that the H became an ever-present sight on town and city streets, especially in France – its low cargo floor and versatile side and rear doors allowing for the easy transportation of goods. The H was replaced in 1958 by an upgraded model, known as the HY, which signalled a series of increases in payload and power. Upgrades continued through the 1960s, with the replacement of the split-screen windscreen by a single screen in 1964 a notable change to the exterior. Although most of the near half million H vans were produced in France, others were built in Belgium, Spain, and the Netherlands, with a small number also assembled in the UK, at Slough.

| SPECIFICATIONS | |
| --- | --- |
| Model | Citroën H |
| Origin | France |
| Built | More than 473,000 |
| Assembly | France, Belgium, Netherlands, Spain |
| Production run | 1948–81 |
| Weight | Varies |
| Payload | 1–1.6 tonnes (1–1.6 tons) |
| Engine | 4 cylinder, 58 bhp (but varies) |
| Transmission | 3-speed manual |
| Maximum speed | 100 km/h (62 mph) |

HOSPITALALWAGEN
BESCHERMING BEVOLKING
LIMBURG ZUID

**Vertical** rear body

**Corrugated** panels

**Citroën ambulance**
The H was used as an ambulance in France and other countries. This van was built in the Netherlands in 1967, from a French "kit". The sign on the side translates as "Protection of the population of Limburg South".

**FRONT VIEW**

**REAR VIEW**

**Citroën advertisement**
A poster from the early 1960s shows two of the practical features – *une porte de côté* ("a side door") and *un plancher bas* ("a low floor") – that Citroën used to promote the versatility of the H as a delivery van.

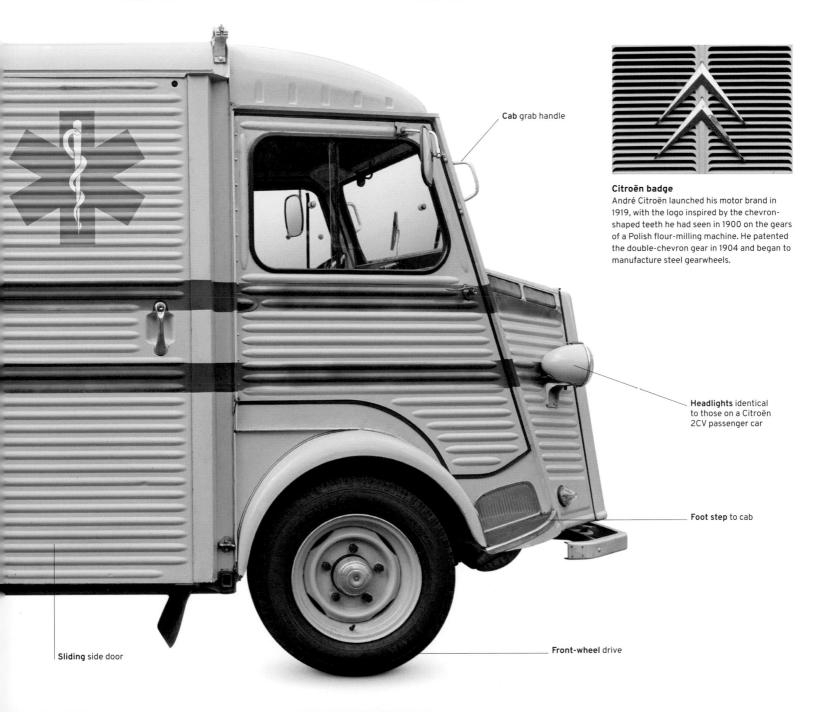

Cab grab handle

Headlights identical to those on a Citroën 2CV passenger car

Foot step to cab

Sliding side door

Front-wheel drive

**Citroën badge**
André Citroën launched his motor brand in 1919, with the logo inspired by the chevron-shaped teeth he had seen in 1900 on the gears of a Polish flour-milling machine. He patented the double-chevron gear in 1904 and began to manufacture steel gearwheels.

## THE EXTERIOR

With outer features that owe little to aerodynamics, it is fair to say the Citroën H was designed for practicality. The flat-sided box shape maximizes the van's carrying capacity, and the ribbed panels – first used on German Junkers aircraft – provide strength at low cost. Loading is extremely flexible. At the rear, a lift-up tailgate and double half-doors provide easy access, while the sliding door at the side of the van is an innovation that would become much copied.

**1.** Radiator cover on front grille to keep engine warm in cold conditions  **2.** Ambulance logo  **3.** Headlight  **4.** Side marker lamp (left-hand side)  **5.** Side marker lamp (right-hand side)  **6.** Rear light unit and electric socket  **7.** Driver/passenger door handle  **8.** Foot step to cab  **9.** Pressed steel wheel hub  **10.** Rear lower door catch  **11.** Fuel filler cap  **12.** Roof attachment mounting

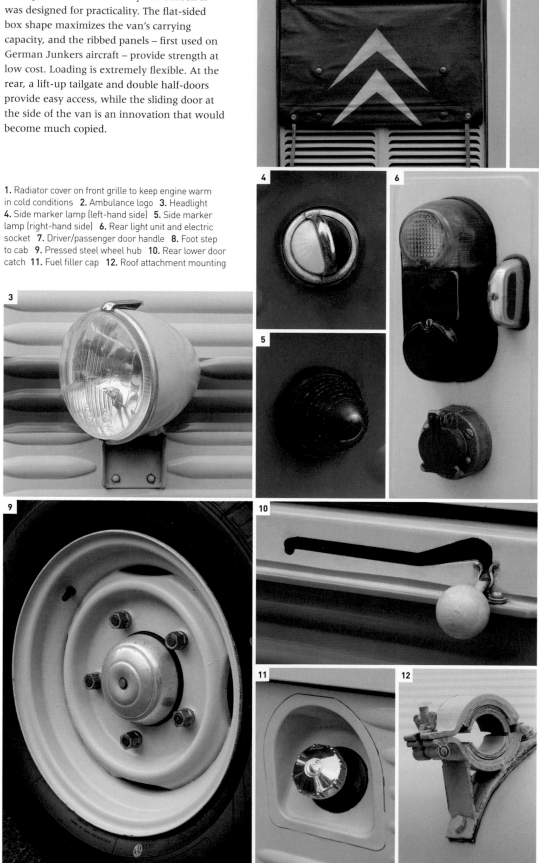

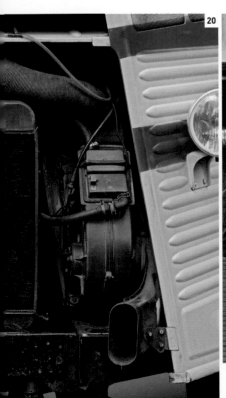

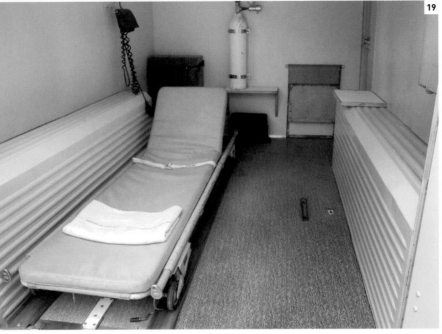

## THE INTERIOR

With no carpets or superfluous pieces of trim, the spartan interior of the H was designed – like the exterior – for low cost and practicality. The reverse-opening front doors, foot steps, and grab handles help to ease entry to the cab, where rubber matting is all that covers the floor. Gear-changing and steering are both heavy but precise. The amount of space created by the low floor and high roof make the H particularly suited to ambulance work.

**13.** Cab interior **14.** Sliding side door handle **15.** Door retaining strap **16.** Interior roof lamp **17.** Windscreen wiper motor **18.** Gear lever and handbrake **19.** Cargo area with ambulance stretcher on runners (taken from similar Citroën ambulance)

## THE ENGINE

Originally used in the Citroën Traction Avant of 1934 – the first mass-produced front-wheel drive car – this is a 1,911 cc three-cylinder petrol engine. The engine is easily accessible – it sits under a metal cover between the front seats – and together with the gearbox, can be removed as one unit for overhauling. Performance is slow, with robustness more of a priority. Various diesel engines were offered from 1961, but the 1,911 cc petrol remained the most popular choice.

**20.** Engine bay with grille lifted **21.** Engine exposed in cab under floor

# Wreckers and Recovery Trucks

The term "wrecker" refers to a truck that is used to move wrecked or broken-down vehicles. In the 1960s, the range of recovery trucks expanded greatly, with 4WD proving itself the must-have platform. Companies such as Holmes in the US and the UK's Harvey Frost were among the few that specialized in manufacturing lifting equipment, or "wrecking gear", that would fit the chassis of recovery trucks, large and small. Any vehicle, from a bus to a tank, could now be recovered swiftly.

△ **Autocar DC75**

**Date** 1960  **Origin** USA

**Engine** 6-cylinder petrol, 335 hp

**Payload** 7.2 tonnes (7.1 tons)

Autocar is one of America's oldest truck companies, producing vehicles for heavy-duty work. This DC75 has the twin-boom Holmes 750 wrecking gear, powered by the truck's engine.

**Blue flag** signifies a lead vehicle in a military convoy

**Adjustable lamp**

**Jerry can** (spare fuel) locker

**Heavy-duty** front bumper in front of radiator grille and lights

△ **Bedford RL Recovery Vehicle**

**Date** 1961  **Origin** UK

**Engine** 6-cylinder petrol, 110 hp

**Payload** 3 tonnes (3 tons)

Bedford's RL was the basis of three medium-sized lorries that served in the British and New Zealand armies from the mid-1950s to early 1970s. The rugged 4×4 chassis proved ideal for fitting out with wrecking gear. The RL could recover other vehicles up to its own weight.

▷ **Jeep FC 170**

**Date** 1963  **Origin** USA

**Engine** 6-cylinder petrol, 105 hp

**Payload** 1.8 tonnes (1.8 tons)

The 4×4, cab-over Jeep 170 featured a power take-off (PTO) recovery winch – that is, a winch powered by the vehicle's drivetrain. Drivers of cab-over wreckers had to take care: overloading could cause the front wheels to lose grip.

**Full-width, counter-balanced hood** makes engine maintenance easier

◁ **Chevrolet C30**

**Date** 1963  **Origin** USA

**Engine** V8 petrol, 175 hp

**Payload** 0.9 tonnes (0.9 tons)

During the 1960s, the C30 was one of the most popular light recovery trucks in the US. Equipped with Holmes 400 wrecking gear and a powerful V8 engine, the C30s were a direct rival to Ford's F-series (see opposite page). One of the later models became the world's fastest tow truck.

◁ **Land Rover 109 Series 2A**

**Date** 1966  **Origin** UK

**Engine** 4-cylinder petrol, 52 hp

**Payload** 750 kg (1,653 lb)

Due to the versatility of the Land Rover, factory-made pickups like this one made an excellent choice for adapting into small breakdown trucks. This example features Harvey Frost recovery gear.

**Steel cables** run from the winch through pulleys to the end of the boom

▽ **AEC Mercury**

**Date** 1969  **Origin** UK

**Engine** 6-cylinder diesel, 151 hp

**Payload** 10.7 tonnes (10.5 tons)

This truck started life as a tractor unit before conversion into a wrecker. It is typical of the sort of truck used by bus companies to recover their vehicles. AEC was a great success story, selling its truck and bus models globally.

△ **Ford F-350**

**Date** 1969  **Origin** USA

**Engine** V8 petrol, 230 hp

**Payload** 0.9 tonnes (0.9 tons)

This 4×4 F-350 has been modified with a twin-wheel rear axle for extra weight-bearing capacity. Mounted on the truck bed is Holmes 440 wrecking gear with folding helper wheels.

**Hydraulic lifting gear** can easily lift articulated trucks

△ **Scania LB110**

**Date** 1969  **Origin** Sweden

**Engine** 6-cylinder turbo-diesel, 11 litre, 271 hp

**Payload** 14 tonnes (13.8 tons)

Scania greatly increased its share of the wrecker market with the LB110, the design being echoed in Scania models for years to come. The version above has hydraulic lifting gear. The air-brakes in the semi-trailer were coupled to the tractor unit.

DAF RB-07-90 carrying
a load of hay, 1956

# Key Manufacturers
# The DAF Story

Today, DAF is one of the biggest-selling truck marques in Europe, with an award-winning range. It is a far cry from the manufacturer's modest origins in an Eindhoven workshop. The Dutch city where it all began now boasts a plant for DAF electric trucks, a nod to the company's legacy of innovation.

**HUBERT JOZEF VAN DOORNE**
– better known as "Hub" – started out in business from a small workshop in 1928, initially making and repairing items, such as ladders, cabinets, and trailers. His work on the latter became the most successful, and in 1932 the company was renamed van Doorne's Aanhangwagenfabriek, and quickly expanded to employ 30 people. Soon after, the less unwieldy acronym DAF came into use.

The business was a partnership between Hub and his brother Wim. Hub concentrated on the engineering and manufacturing side while Wim oversaw the administration and finances. In those early years, Hub proved himself to be an innovator

DAF 2200, 1970s

with trailers. In 1931, he applied for DAF's first patent for a semi-trailer coupling that automatically lifted the support legs when the trailer was being coupled to a tractor. A revolutionary idea, this meant the driver no longer had to get out of the cab to couple it.

DAF's first foray into trucks came in 1934, when Hub, along with a lieutenant in the Dutch Army, Mr van der Trappen, developed a "Trado conversion set" that could turn a two-axle truck into a six-wheeler with four driven wheels in just a few hours. However, it was not until after World War II that DAF began to manufacture trucks – mainly for the Dutch military – using components from other companies. Their first model was the 3-tonne (3-ton) A30, which launched in 1949 with an imported Hercules 91-hp petrol engine. The versatile model proved successful and was upgraded several times in the coming years.

DAF briefly moved into car manufacturing in the 1950s, before selling the division to Volvo in 1975. The company also began developing its own engines in the 1950s after initially using petrol and

diesel units produced by other manufacturers. DAF's first engine came out in 1959 and was a leap forwards, not only for the company but the industry, as one of Europe's first turbocharged diesel engines.

In 1962, as longer-distance deliveries became more common, DAF launched the 2600, the first European truck that had a sleeper cab, with a bunk and proper mattress fitted. This was a revolution in comfort and saved drivers having to find lodgings or sleeping across a seat.

DAF was again ahead of the competition when it became the first in the haulage industry to introduce turbo intercooling in

**DAF A-60 tractor trailer, 1950**
DAF trucks have always been popular fleet trucks. Their cabs are renowned for driver comfort, which became more important as long-distance deliveries increased.

1973. At the time, this process, which cools down the compressed air generated by the turbocharger, was designed to deliver better engine performance and reduce fuel consumption. It also became important for realizing cleaner exhaust gases in later years. The process was further refined to make greater efficiencies and power gains in 1985 with the introduction of ATi, or Advanced Turbo Intercooling.

Also in 1985, DAF launched its SpaceCab, with even more space for drivers than in previous models, enabling "trampers" – those who worked away all week – to live in more comfort. This was followed in 1994 by the Super SpaceCab, which set the benchmark for cab space for decades afterwards.

In 1962, DAF **launched** the 2600, the **first European truck** that had a **sleeper cab,** this was a **revolution** in **driver comfort.**

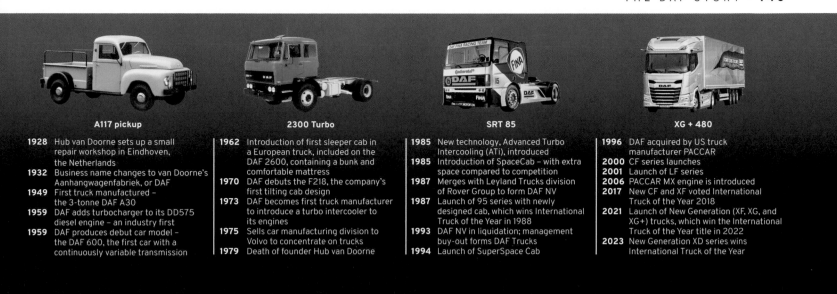

**A117 pickup**

**2300 Turbo**

**SRT 85**

**XG + 480**

| | |
|---|---|
| **1928** | Hub van Doorne sets up a small repair workshop in Eindhoven, the Netherlands |
| **1932** | Business name changes to van Doorne's Aanhangwagenfabriek, or DAF |
| **1949** | First truck manufactured – the 3-tonne DAF A30 |
| **1959** | DAF adds turbocharger to its DD575 diesel engine – an industry first |
| **1959** | DAF produces debut car model – the DAF 600, the first car with a continuously variable transmission |

| | |
|---|---|
| **1962** | Introduction of first sleeper cab in a European truck, included on the DAF 2600, containing a bunk and comfortable mattress |
| **1970** | DAF debuts the F218, the company's first tilting cab design |
| **1973** | DAF becomes first truck manufacturer to introduce a turbo intercooler to its engines |
| **1975** | Sells car manufacturing division to Volvo to concentrate on trucks |
| **1979** | Death of founder Hub van Doorne |

| | |
|---|---|
| **1985** | New technology, Advanced Turbo Intercooling (ATi), introduced |
| **1985** | Introduction of SpaceCab – with extra space compared to competition |
| **1987** | Merges with Leyland Trucks division of Rover Group to form DAF NV |
| **1987** | Launch of 95 series with newly designed cab, which wins International Truck of the Year in 1988 |
| **1993** | DAF NV in liquidation; management buy-out forms DAF Trucks |
| **1994** | Launch of SuperSpace Cab |

| | |
|---|---|
| **1996** | DAF acquired by US truck manufacturer PACCAR |
| **2000** | CF series launches |
| **2001** | Launch of LF series |
| **2006** | PACCAR MX engine is introduced |
| **2017** | New CF and XF voted International Truck of the Year 2018 |
| **2021** | Launch of New Generation (XF, XG, and XG+) trucks, which win the International Truck of the Year title in 2022 |
| **2023** | New Generation XD series wins International Truck of the Year |

The company has been reshaped due to competition with its rivals over the years. In 1987, it merged with the Leyland Trucks arm of the Rover Group to form DAF BV. That meant that trucks in the UK were badged as Leyland DAF, but DAF in all other markets. This also gave DAF a UK manufacturing plant, which is still in use today. However, in February 1993, after a downturn in sales, the company was placed in receivership and sold in three management buy-outs to create DAF Trucks, based in the Netherlands, as well as Leyland Trucks and LDV Group, both based in the UK. DAF and Leyland Trucks were later bought by PACCAR and are still part of the US company.

Today, DAF continues to innovate. In November 2021, it launched its New Generation of long distance trucks. With the largest cabs in their class and improved aerodynamics to boost fuel economy, the New Generation scooped the 2021 International Truck of the Year Award. DAF has also invested heavily in electric trucks in recent years to be at the forefront of the shift to zero-emissions transport. The company had already developed three generations of fully electric versions of its current vehicle line-up, when in 2023 it opened a dedicated electric truck manufacturing facility in Eindhoven – the Dutch city where DAF was founded.

**Hard worker**
DAF trucks, such as those in the Vocational Series, have been fixtures in heavy industry and construction for decades, thanks to their comfort, durability, and reliability.

# Advancements in Cabs

Before the 1960s, aesthetics were often an afterthought in the design of specialist trucks, but during this decade truck designers began to blend practicality with style. Commercial manufacturers began taking a cue from the increasingly refined styling of the car market. In order to meet new legislation on fuel efficiency, manufacturers were tasked with designing more aerodynamic trucks that appealed to their customers. As Cold War tensions grew during this decade, many specialist military trucks were introduced, taking advantage of the latest technologies.

**Roof-mounted** search light

**▽ Alvis Stalwart**

| | | | |
|---|---|---|---|
| **Date** 1960 | **Origin** UK | | |
| **Engine** V8 petrol, 185 bhp | | | |
| **Payload** 5 tonnes (4.9 tons) | | | |

The 6x6 amphibious Stalwart was developed to transport cargo across water and rough terrain. Some were fitted with Atlas hydraulic cranes. Its watertight cab featured a central driving position and was accessed by two roof hatches.

**▷ Willys Jeep FC-150**

| | |
|---|---|
| **Date** 1960 | **Origin** USA |
| **Engine** 4-cylinder petrol, 72 hp | |
| **Payload** 780 kg (1,720 lb) | |

Jeep's Forward Control (FC) series was introduced in 1956. This new 4x4 pickup had a cab-over and a generous payload for its very small wheelbase of just 206 cm (81 in). The bulk of these had standard pickup bodies, but later models included vans.

**Rear** quarter light window

**Substantial door mirrors** have a secondary mirror pointing towards the ground

**▽ International Sightliner**

| | |
|---|---|
| **Date** 1960 | **Origin** USA |
| **Engine** V8 petrol, 256 hp | |
| **Payload** 18 tonnes (17.7 tons) | |

To reduce the size of the cab to meet new regulations, the Sightliner was fitted with a short V8 engine and had extra windscreens on the foot wells. These added windows, however, acted as magnifying glasses and burnt drivers' legs in hot weather.

**Cab** was developed by Leyland Motors in 1965

R.C. JEFFREY & SONS

Transport Contractors
PEBWORTH
STRATFORD UPON AVON
Phone 2131 534

**Fuel tank**

**Battery location/** rope and sheet storage

▽ **Chevrolet C10 Panel Van**

**Date** 1962  **Origin** USA

**Engine** 6-cylinder petrol, 135 hp

**Payload** 500 kg (1,100 lb)

The new 1960 Chevy cabs featured a wrap-around windscreen, which increased drivers' visibility. Some drivers much preferred the later models as the door pillar itself presented a blindspot. This model has modern alloy wheels fitted.

**Door pillar** caused a blindspot

**TECHNOLOGY**

# Truck aerodynamics

Aerodynamics were taken seriously in the 1960s in an effort to increase speed, reduce fuel consumption, and improve efficiency. As trucks became boxier, they needed to be less resistant to airflow. Roof-mounted wind deflectors, or "breakers", helped direct the incoming airflow up and over the cab, and away from the trailer behind. Eventually, the cavity within the breaker was put to use, by inserting sleeper beds, accessed through the cab roof. Fibreglass mouldings were fitted to the cab's angled corners, and the bumpers were lowered, pushing the drag of the airflow away from under the truck. These developments all steered the evolution of truck aerodynamics in subsequent decades.

**NASA aerodynamic truck** This cab-over truck was enveloped in light-weight aluminium panels by NASA to conduct their own experiments into vehicle drag.

**Cab roof storage** was popular in Europe

△ **Dodge W200 Power Wagon**

**Date** 1962  **Origin** USA

**Engine** Petrol, 94 hp

**Payload** 750 kg (1,650 lb)

The cab of the Power Wagon was redesigned in 1962 and remains the standard configuration today. The chassis was lengthened to accommodate the bigger crew cab, with bench seating front and rear, while keeping a practical loading area.

**Access panel** to engine

◁ **Scania 76**

**Date** 1967  **Origin** Sweden

**Engine** Diesel, 190 hp

**Payload** 10 tonnes (9.8 tons)

The first Scania truck with a cab-over, the 76 model set a new trend for Scania. Legislation at the time to reduce the length of articulated trucks resulted in more cab-over designs. The 76 was popular around Europe, and was the first Scania introduced into the UK market.

The 8x4 chassis was the longest AEC available

◁ **AEC Mammoth Major**

**Date** 1969  **Origin** UK

**Engine** 6-cylinder diesel, 212 bhp

**Payload** 15 tonnes (14.8 tons)

The flat-fronted "Ergomatic" cab of the Major was a forerunner to the 1970s era of design. The tilting cab had more glass than before, improving driver visibility and letting more light in. The cab was very short, enabling the full potential of the chassis to be used, such as on this 8x4 example.

# Coming Through

Road trains are a common sight in Australia. The country has the longest and heaviest road-legal road trains in the world, weighing up to 200 tonnes (197 tons) and measuring up to 53 m (175 ft) in length. These transport a wide range of goods across the country, including fuel, mineral ores, and general freight. Road trains are used as they are more cost-effective than other forms of transport.

## VITAL LIFELINE

Road trains have existed in Australia since the mid-19th century, when traction engines were used to pull trailers. Using trucks for road trains first developed in the 1930s and '40s. Bush mechanic Kurt Johannsen is credited with inventing the modern road train, when he combined a Diamond-T tank carrier and two self-tracking trailers. Road trains have helped Australia's more remote spots to develop economically, and some areas now rely on the regular service they provide to get key provisions. Other countries, such as the USA, Mexico, and Finland, also have road trains, but none are as large as their Australian counterparts.

**A MACK Titan truck with a triple road train** travels through the Australian outback. This heavy road-train model first launched in Australia in 1995.

# The Road Ahead

In the 1960s, the modern road networks that are taken for granted today were beginning to take shape. Following on from earlier road-building projects in Europe, the decade saw the construction of America's new Interstate Highway System, as well as the UK's first motorways.

In the 1950s and 1960s, advances in design and technology allowed trucks to go further and faster. Larger trucks with more powerful turbocharged engines were able to haul bigger loads and travel much greater distances. To make the most of these new developments, it was vital for road infrastructure to keep pace, and the 1960s saw a rapid expansion in the construction of motorway networks linking major towns and cities.

The concept of the motorway was not a new idea: in Italy, a stretch of dual carriageway road between Milan and Varese had opened to the public back in 1924. Now known as the A8, it is considered to be Europe's first motorway. Furthermore, construction of the Bonn–Cologne Autobahn had begun in Germany in 1929, with the country's famous road network expanding throughout the 1930s.

The difference in the postwar period was the scale and speed of these road infrastructure projects.

**Building the Brenner Autobahn, 1968**
Linking Innsbruck in Austria to Modena in Italy, the Brenner Autobahn was the first motorway to cross the Alps. The viaduct above traverses the Obernberg Valley in Austria's Tyrol region.

**M1 motorway, UK, 1965**
In its early years, the M1, the UK's first full motorway, had no speed limits, crash barriers, or lights, and soft shoulders rather than hard.

# Motorway construction in the UK began in 1956, with the first 1,600 km (1,000 miles) completed by 1969.

In 1961, France opened the 63-km (39-mile-) long Estérel–Côte d'Azur toll motorway. By 1964, Germany's growing autobahn system extended over 3,000 km (1,900 miles). The UK made a more tentative start in 1958 with the 13-km (8-mile) Preston Bypass. A year later, the initial section of the M1, the country's first full motorway, opened to the public.

**From ship to trailer, 1968**
Goods containers that can be used by different transport modes – ship, rail, or truck – cut delivery costs and time. Cranes can transfer containers directly from ship to truck.

## From sea to shining sea

Perhaps the most famous symbol of mid-20th century highway expansion can be found in America, with the Dwight D. Eisenhower National System of Interstate and Defense Highways (more commonly known as the Interstate Highway System).

As a young Lieutenant Colonel, Eisenhower was involved in the US Army's first Transcontinental Motor Convoy, a military expedition across the United States that highlighted the huge inconsistencies in road quality. During World War II, while acting as Supreme Commander of the Allied Expeditionary Force, Eisenhower was impressed by the German autobahn system, realizing the importance of reliable road networks in maintaining a steady supply of goods.

In 1956, during his first term in office as US president, Eisenhower signed the National Interstate and Defense Highways Act. Construction would continue throughout the 1960s, with Nebraska becoming the first state to complete all of its Interstate Highways in 1974.

## Going global

The 1960s saw goods distribution increasingly shifting away from railroads and towards trucking, in part due to bigger, better trucks and new highways. But other infrastructure developments also contributed to turning road freight into a global operation.

Key among these was the arrival of the modern shipping container, which was pioneered by American truck driver and entrepreneur Malcolm McLean in 1956. By the 1960s, his idea had evolved into the standardized intermodal shipping

container system seen today. The system allows time-sensitive goods to be transported much more quickly, with containers lifted directly onto truck trailers without the need to unload them first.

The decade also saw more frequent ferry crossings in Europe and an increase in bridge-building projects, helping to move trucks more quickly and in greater numbers. Major engineering schemes, including the Mont Blanc Tunnel, also pushed ahead, establishing newer, more efficient trucking routes. The 11.6-km (7.2-mile) tunnel opened in 1965, linking France and Italy beneath the highest mountain in the Alps. While it may be less picturesque than the traditional winding mountain route, the tunnel saves time, fuel, and money for the trucking industry.

**TALKING POINT**

## Giving Truck Drivers a Break

As journeys became longer, it was important to make sure the needs of drivers were still being met during lengthy periods on the road. The first purpose-built truck stop in the US opened in Nebraska in 1948, offering food, fuel, and rest for tired truckers. As the highways expanded, the truck stops proliferated. The advent of the Interstate Highways saw franchises beginning to take over from the traditional "Mom and Pop" stores, with larger chains adding retail units, food courts, showers, and amusement arcades.

Today, there are more than 40,000 truck stops in the US. In the UK, many motorway service stations have truck areas, while in Germany and Austria, state-owned service stations work alongside privately owned Autohofs.

**Billboard for diesel** at a truck stop on Interstate 80, Iowa

# Car Transporters

Following the creation of Formula 1 and Nascar in the 1940s and 50s, motor racing became increasingly popular. Car manufacturers knew that success on the track inspired sales –"Win on a Sunday, sell on a Monday" – so were keen to compete. This meant having specialized trucks to transport their cars to the tracks, as well as having other support vehicles. Some transporters were adapted from coach platforms, as seen in the converted coach that carries Minis in the 1969 movie *The Italian Job*. In addition to trucks that ferried racing cars between events, there were also those that transported road-going cars from the factory to the customer. These were typically tractor and trailer units, or medium-sized trucks.

**△ Fiat 682-R Ferrari Transporter**

| | |
|---|---|
| **Date** 1957 | **Origin** Italy |
| **Engine** 6-cylinder diesel, 140 hp | |
| **Payload** 14.2 tonnes (14 tons) | |

Fiat built this vehicle for the Ferrari race team in 1957, and it was used until 1969. Two cars are carried on the top deck and one on the bottom. The remaining space was for crew and spare parts, doubling as a repair workshop.

**Double side doors** allow access for loading and unloading

**▷ BMC Transporter**

| | |
|---|---|
| **Date** 1959 | **Origin** UK |
| **Engine** 6-cylinder diesel, 89 hp | |
| **Payload** 5.1 tonnes (5 tons) | |

Designed by Pininfarina of Italy and built by Marshall's of Cambridge, this truck was one of 20 built as travelling workshops. They were used to train mechanics who would soon be working on the newly launched Austin Mini motorcar.

**All-aluminium** bodywork

**▽ Commer TS3 Three-Car Transporter**

| | |
|---|---|
| **Date** 1960 | **Origin** UK |
| **Engine** 3-cylinder, 2-stroke diesel, 105 bhp | |
| **Payload** 11.6 tonnes (11.5 tons) | |

Built to carry the Jaguar D-types of the Ecurie Ecosse race team, this double-decker had a workshop behind the crew cab, allowing access to the underside of a car on the top deck. After refurbishment, it sold for £1.79 million in 2013.

**△ Ford F-350 Ramp Truck**

| | |
|---|---|
| **Date** 1963 | **Origin** USA |
| **Engine** V8 petrol, 400 hp | |
| **Payload** 3.5 tonnes (3.5 tons) | |

This converted F-350 has a lengthened chassis, a custom-built ramp bed, and side lockers. It also has a fifth-wheel coupling, enabling it to haul multiple cars on a trailer if necessary.

**Flashing beacon** alerts others to a potential hazard

**▽ Mercedes-Benz L 405**

| | |
|---|---|
| **Date** 1963 | **Origin** Germany |
| **Engine** 4-cylinder diesel, 50 hp | |
| **Payload** 1.8 tonnes (1.8 tons) | |

The L series chassis was configured for vans, light trucks, buses, and fire engines, but no version was simpler than that of this recovery truck, which spent its working life in Italy.

△ **Bedford TK**

| | |
|---|---|
| Date 1965 | Origin UK |
| Engine 6-cylinder diesel, 98 hp | |
| Payload 11.8 tonnes (11.6 tons) | |

The TK was the UK's foremost transporter during the 1960s and 1970s, delivering new cars from factories to dealerships and ports all over the country. This five-car transporter had a hydraulically operated top deck and ramps that slide out from the rear.

## Citroën HY

The Citroën HY was a light French commercial vehicle first produced in 1958. Some long-wheelbase versions of the HY van were adapted for carrying cars. Although not produced in large numbers, HY car transporters were seen in garages across France. The HY transporter made use of the Citroën corrugated panel work, which reduced the transporter's weight. The spartan interior needed little upkeep, contributing to vehicle's longevity. The transporter was far from speedy, but this did not prove a disadvantage, since the emphasis was on local deliveries.

**This Citroën HY car transporter** is loaded with a Citroën GS car. The example shown here is purpose-built, but some standard HY pickups were also adapted to transport small cars.

**Storage space** used for wheels and tyres

⊲ **Ford F-350 Race Truck**

| | |
|---|---|
| Date 1966 | Origin USA |
| Engine V8 petrol, 320 hp | |
| Payload 3.5 tonnes (3.4 tons) | |

Most pickup-based transporters had simple, ramp-based bodies. This F-350, however, has been designed to accommodate storage for tools and spares, as well as rest space for the race crew.

shelby american, inc.
6501 w. imperial hwy.
el segundo california

FORD G.T.    COBRA    SHELBY G.T. 350

**Top deck** could accommodate two cars

**Upper ramp** was raised and lowered by a hydraulic ram

ECURIE ECOSSE

GRAND PRIX SUISSE
www.gp-suisse.ch

1970s
# LIVING THE DREAM

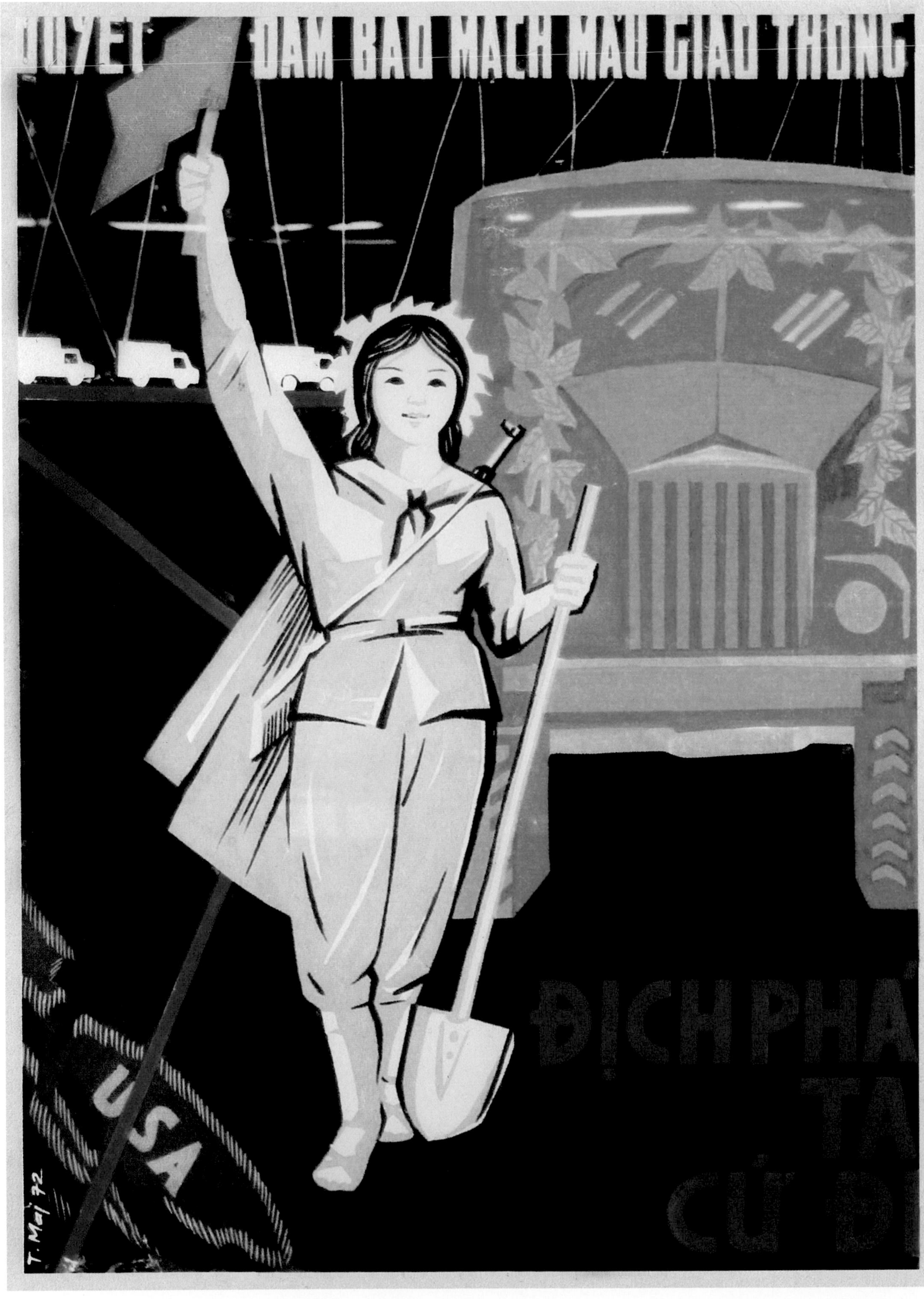
ĐẢM BẢO MẠCH MÁU GIAO THÔNG
USA
T. Mai 72

# LIVING THE DREAM

**The 1970s brought trucks to Hollywood with the** release of such movies as *Convoy* (1978) and *Smokey and the Bandit* (1977). The big rig had become cool – bedroom posters of Kenworths replaced those of muscle cars. Trucking and the trucking lifestyle had become so popular that songs, even bands, were dedicated to them. Owner-drivers from all nations took pride in customizing their trucks with artwork and lights.

△ **A truck in France, 1978**
A converted truck being used as a food stall at a market in Saintes-Maries-de-la-Mer's church square.

In Europe during this time, while the British motor industry was being crippled by strikes at British Leyland, Volvo of Sweden introduced their new top-of-the-line "Globetrotter" series, a benchmark for driver refinement. The F88, with its sleeper cab and power steering, was a game-changer for long-haul journeys. Meanwhile, in Africa, the construction of the 4,500-km (2,800-mile) Trans-Sahara Highway across the Sahara helped open up the African continent to road freight.

Life improved for the drivers, too. Radios in trucks had long helped pass the time, but now a new craze gripped trucking communities – the CB radio (Citizens Band), immortalized in American movies. If a lonely trucker wanted friendly advice, it was only ever a call sign away. The ratchet strap also became the trucker's best friend during this decade. Securing loads used to be time consuming and frustrating, but ratchet straps offered a quick, easy, and safe method for doing so. They meant that a driver could focus on the drive without worrying that their load was not secure.

△ **A fuel truck in Cambodia, 1975**
An operator sits alongside his medium-duty Chevrolet C60 truck as a petrol shortage grips Phnom Penh, Cambodia.

**In 1975,** the **USSR** produced **872,000 trucks and buses,** compared with **1,634,000** made in the **United States.**

## Key events

▷ **1970** Bedford MK is introduced for the British armed forces and other nations. Based on the TK, the new 4x4 truck becomes the workhorse of the services.

▷ **1973** Chevrolet Blazer MK2 launches. It sets a signature style for the Blazer line that remains largely unaltered for two decades.

▷ **1973** During the Suez Crisis, political issues spurred a dramatic rise in the cost of oil, and with it, large-scale fuel shortages that had a knock-on effect on global trucking.

▷ **1975** IVECO launches in Italy to incorporate a new truck range.

△ **IVECO launches new range of trucks, 1975**
Five European truck manufacturers set out to lead the heavy-duty market by forming IVECO, merging their existing lines into a comprehensive range in 1975.

▷ **1977** *Smokey and the Bandit* is released, starring Burt Reynolds and a 1974 Kenworth W900 series truck.

▷ **1977** Lada Niva goes on sale in Russia and Europe. The 4x4 chassis has many uses, from pickups to fire engines.

▷ **1977** Honda Acty, the tiny Japanese pickup truck with a 545cc engine becomes an international hit.

▷ **1979** Bigfoot monster truck of St Louis (a Ford F25) becomes famous, spawning a series of stadium events.

▷ **1979** Volvo G89 truck becomes popular for long-distance journeys.

▷ **1979** The first official truck race starts at the Atlanta Speedway.

◁ **Vietnamese patriotic poster** from the 1970s, which reads, "Ensure the passage of truck convoys".

# Popular European Tractor Units

The 1970s were a time of change in European haulage. Long-distance trucking across the continent was becoming commonplace, and loads were becoming larger. As a result, trucks were changing too: the decade saw the emergence of sleeper cabs for drivers – and greater cab comfort in general – and engines with ever-greater horsepower. In addition, cab-over trucks, as opposed to "conventional" models with bonnets, were becoming more popular – a trend that endures in Europe to this day. Manufacturers were also looking more to export markets, with Volvo, Scania, Mercedes, and others experiencing rapid expansion abroad.

## ▽ Volvo F88

**Date** 1972  **Origin** Sweden

**Engine** 6-cylinder, 200–312 hp

**Payload** 37–52 tonnes (36–51 tons)

The first of Volvo's System 8 trucks, the F88 was an export success, cementing Volvo's reputation for reliable cab-over trucks. It sold over 60,000 F88s and F89s (its sister model) from 1965 to 1977.

Cab sits over the engine

## ▽ Scania 140

**Date** 1972  **Origin** Sweden

**Engine** V8, 350 hp

**Payload** 32 tonnes (31 tons) GVW

Scania's 140 was the first model to feature its new V8 engine, which gave extra power at all speeds. The 140 raised Scania's profile, winning fans in the UK and Europe for its performance on long-distance haulage routes.

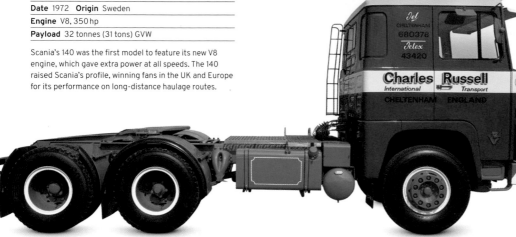

## ▽ Saviem SM240

**Date** 1975  **Origin** France

**Engine** 6-cylinder, 202 hp

**Payload** 32 tonnes (31 tons) GVW

This was one of many models to have the European cab-over design that was developed as part of Saviem's partnership with MAN. This collaboration also saw the companies' trucks use the same engines.

## ▷ Atkinson Borderer Mark II

**Date** 1975  **Origin** UK

**Engine** V8, 240 hp

**Payload** 32 tonnes (31 tons) GVW

The Atkinson Borderer was launched in 1970, shortly before the company became Seddon-Atkinson. Appreciated for its simple build, the Borderer remained a fixture of long-haul transport in the UK for many years.

## TALKING POINT

## Saviem SM8

The SM (Saviem–MAN) medium-duty truck range was produced by French manufacturer Saviem, in partnership with MAN of Germany, between 1967 and 1975. Named for its 8-tonne (8-ton) payload, the rear-wheel drive SM8 was equipped with a straight-six engine. Known for its versatility, it was available as a panel truck, flat-bed, or tanker. A 4×4 version was used by the French army.

**Saviem SM8 in Côte d'Ivoire** The SM8 was particularly popular in African countries that had close ties with France.

Tilting cab

## ▽ Volvo F86

**Date** 1976  **Origin** Sweden

**Engine** 6-cylinder, 150–210 hp

**Payload** 16.5 tonnes (16 tons) GVW

F86 production lasted from 1965 to 1979. Efforts to expand overseas led Volvo to set up assembly in Scotland for UK sales. A stalwart of the road, the F86 soon became the UK's most popular truck.

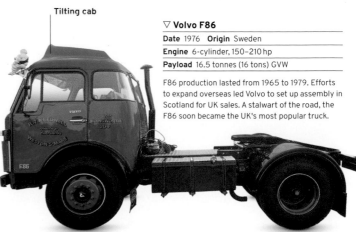

**Shipping container**
was a common load
for the F88

**Coiled cables**
("susies") supply
electrical power
and air from
tractor to trailer

**Extendable legs** support
trailer when it is uncoupled

▷ **Volvo N12**

| Date | 1976 | Origin | Sweden |
|---|---|---|---|
| Engine | 6-cylinder, 367 hp | | |
| Payload | 32 tonnes (31 tons) GVW | | |

The N12 had four wheels at the rear, which
gave great on- and off-road adhesion. It
sold well in Scandinavia, UK, US, South
America, and Australia, where it was often
used to pull road trains.

**Day cab,** with no sleeping
berth (crew and sleeper cab
options were also available)

▽ **Mercedes 1622**

| Date | 1976 | Origin | Germany |
|---|---|---|---|
| Engine | 6-cylinder, 220 hp | | |
| Payload | 16 tonnes (16 tons) | | |

The medium-weight 1622 was popular across
Europe for many years in the 1970s and 1980s.
A versatile model, it was often configured as a
flatbed truck or a tipper. The 1622 became
known for its low fuel consumption.

◁ **ERF B Series**

| Date | 1977 | Origin | UK |
|---|---|---|---|
| Engine | V8, 244–261hp | | |
| Payload | 32 tonnes (31 tons) GVW | | |

The B Series truck had an SP (steel/plastic)
cab. The cab tilted for easy engine access –
a first for an ERF production model. It was
offered with Gardner, Cummins, or Rolls
Royce engine options, and day-cab or
sleeper-cab configurations.

**Fifth-wheel** coupling

# Scammell Contractor

Designed and built under a partnership between British Leyland and Scammell, the 244-tonne (240-ton) Scammell Contractor pulled the heaviest loads on the road. Often referred to as "ballast tractors", these powerful vehicles were designed to ensure they had the traction required to haul almost anything that could fit on the road, from trains and aeroplanes, to ships.

**THE CONTRACTOR LOOKED** as if it could carry a load in its rear body, but this was only designed to hold ballast weight to maintain traction above the rear wheels. Like all ballast tractors, this weight prevents wheelspin and improves traction when hauling heavy loads on trailers. Nicknamed "Intruder", this example remained in service with Abnormal Load Engineering until 2002, after which it was completely stripped down and rebuilt from the chassis up. Toymaker Corgi even produced a diecast model of it, complete with the type of heavy-haulage trailer it would have used in service.

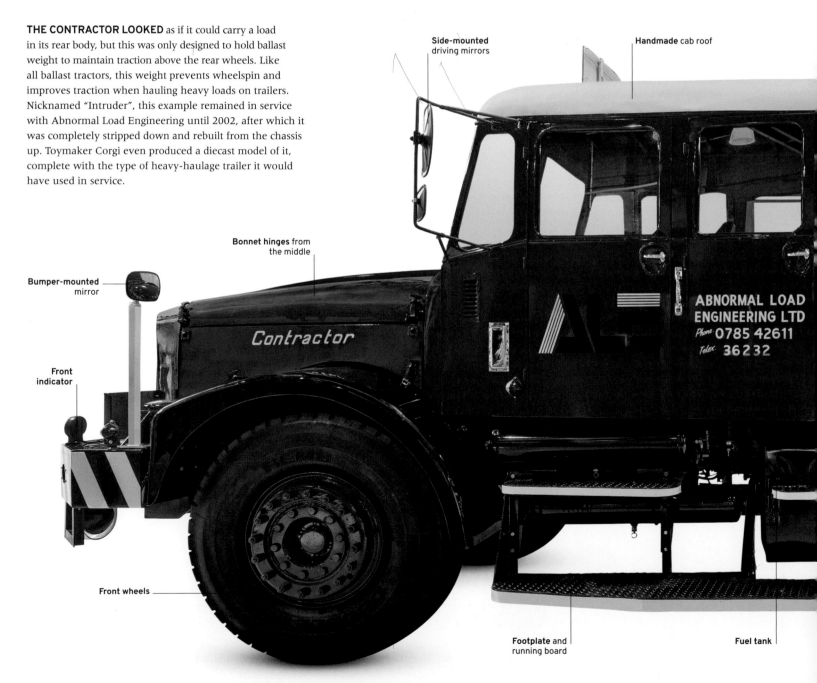

Side-mounted driving mirrors

Handmade cab roof

Bonnet hinges from the middle

Bumper-mounted mirror

Front indicator

ABNORMAL LOAD ENGINEERING LTD
Phone: 0785 42611
Telex: 36232

Contractor

Front wheels

Footplate and running board

Fuel tank

**FRONT VIEW**

**REAR VIEW**

**Rear worklamp**

**Rope belts** for
tying down cover

**Exhaust pipe**

### Heavy lifter

The trailers these ballast tractors towed were so
large that they required a separate cab at the rear
with their own steering. The rear axle set had a total
loading capacity of 41 tonnes (40 tons). An upgraded
version was supplied as a tank transporter to the
British and Australian armies.

| SPECIFICATIONS | |
|---|---|
| Model | 1976 Scammell Contractor |
| Origin | UK |
| Assembly | Watford, UK |
| Production run | Unknown |
| Weight | 37 tonnes (36 tons) |
| Payload | 244-tonne (240-ton) gross weight capacity |
| Engine | 6-cylinder turbo diesel, 14 litre, 335 bhp |
| Transmission | RV 30 semi-automatic air shift |
| Maximum speed | 58 km/h (36 mph) |

A.L.E. STAFFORD 42611

## THE EXTERIOR

Designers of the 6x4 Contractor had a US market in mind, giving it a large double cab that lends itself to the heavy haulage sector, or "Prime Movers" as it is known in the US. The cab itself is constructed around a steel and aluminium frame with a mixture of aluminium outer panels.

**1.** Scammell radiator cast badge   **2.** Name given to this truck by its owner   **3.** Loading coupling   **4.** Externally-mounted air filter   **5.** Air intake pipe from air filter   **6.** Cab vents   **7.** Cab door handle and grab handle   **8.** Rear-axle leaf springs   **9.** Rear wheel   **10.** Air coupling for brakes   **11.** Air feed connecter line

## THE INTERIOR

More in keeping with a railway locomotive than a truck due to the size of the vehicle itself and the extreme loads it was built to move, the cab interior has an industrial feel. Most of the controls are air operated. The cab was designed to allow engineers to jump out at quick notice – often necessary when crawling through towns towing vast overhanging objects. The driving position provides commanding visibility over the large bonnet, with the huge steering wheel installed at an acute angle to help make the task of constantly turning it as manageable as possible.

**12.** Driving controls and dashboard   **13.** Rear crew seat   **14.** Cab roof lamp fixing
**15.** Dash-mounted parking brake lever   **16.** Air-operated gear lever

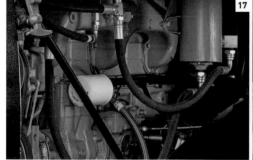

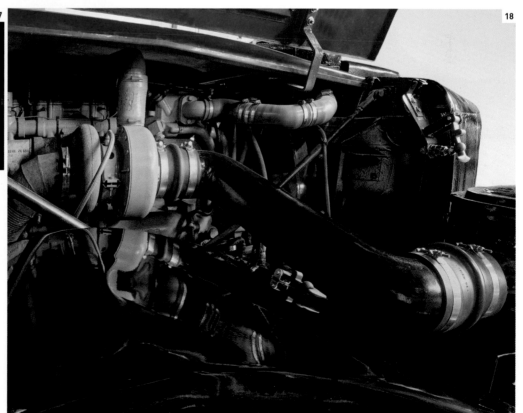

## THE ENGINE

One of the most successful modern diesel engines ever built, the Cummins 335 was known for its reliability and endurance. It was able to clock up over 2.5 million km (1.5 million miles) before needing an overhaul. This turbocharged engine was also used in a wide variety of US trucks. It featured a system that could recycle unburned fuel from the combustion chamber. The fuel heater helped it start in cold climates.

**17.** Cylindrical oil and fuel filters at base of engine
**18.** Turbocharger that draws air into the engine inlet manifold in order to increase power

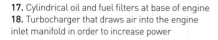

# Military Muscle

During the Cold War era (1947–1991), behemoths such as the MAZ-537 (below) were a regular sight in the USSR. Each year, massive 16-wheeled trucks would haul enormous warheads through Red Square as part of Moscow's November parade. It was a golden opportunity for the Communist Party to show off its military strength both to its people and to rival nations.

## ERA OF INNOVATION

As the Cold War led to the proliferation of huge intercontinental ballistic missiles (ICBMs), the need arose for ever-larger trucks to carry them – and ones that could also cope with rough, often muddy terrain, and harsh Soviet winters. This triggered an era of innovation among Soviet truck engineers. Previously, trucks in the USSR had mostly been based on American models of the 1930s, but domestic truck manufacturers such as ZiL, Ural, and MAZ now began producing impressive new truck designs themselves. They included 8×8 and even 16×16 wheeled configurations – some of the largest trucks ever built. The boom in new truck design extended to many other forms of military transport, from troop carriers to rocket launchers.

**MAZ-537 transporter units** crawl slowly along with their ICBM payloads at the November parade in Moscow, 1977. A large banner depicting communist workers adorns the State Historical Museum behind.

# American Big Rigs

The 1970s was a pivotal decade for the development and expansion of US semi trucks, or "rigs". By the middle of the decade, over one million of them were heading out across America. The cab became a home away from home, allowing drivers to spend more time on the road, which led to the rise of trucker "communties". Featuring chromed stacks (exhausts) – that were more cosmetic than practical – and often richly painted, rigs began featuring on bedroom pin-up posters. Manufacturers released a massive variety of these trucks, catering to the needs of every sector, and maximum payloads varied depending on the exact specification requested by the owner.

▽ **International F-210D**

**Date** 1970 **Origin** USA

**Engine** Cummins NH, 200 hp / 250 hp

**Payload** 27 tonnes (26.6 tons) max GVW

This rig's simple and durable design meant it was easy to maintain. The workhorses of the US Air Force and the US as a whole, they were also exported in large numbers to Australia and New Zealand, where they were ordered as fleets.

**Hopper trailer** is filled from the top and empties at the bottom

▷ **Ford W9000**

**Date** 1971 **Origin** USA

**Engine** Detroit 318 diesel

**Payload** Variable

Ford is not generally well known for building semis, but the W9000 cab-over model was a cost-effective answer to new legislation brought in to regulate the lengths of semi trucks in some states.

**Sleeper cab** has air-powered wiper motors

**Tri-axle** means this truck can pull heavier loads

△ **Mack DM 863 SX**

**Date** 1971 **Origin** USA

**Engine** Cummins 6-cylinder turbo diesel

**Payload** Variable

With a good reputation for being functional and rugged, this truck has a "helper axle" that is lowered to spread the weight of a heavy load. Like all Mack trucks, this was a versatile model, used for long hauls between US states, and for tough industrial work, on and off the highway.

▷ **Peterbilt 359**

**Date** 1971 **Origin** USA

**Engine** Caterpillar V turbo diesel

**Payload** 23 tonnes (22.6 tons)

One of the most cherished of all the American big rigs, the 359 featured classic styling with vertical exhaust stacks, a chrome air breather, and a long-lasting aluminium cab. Its air conditioning, large engine, and transmission options were just a few of the ingredients that made the 359 one of America's best-loved trucks.

▷ **Autocar DC9364**

**Date** 1972 **Origin** USA

**Engine** Cummins NTC turbo diesel, 335 hp

**Payload** Variable

Established in 1897, Autocar is the oldest American truck manufacturer. This heavy-duty tractor was converted from a twin rear axle (that could haul more) to a single during its restoration.

**Chrome** radiator surround

**External** air filter

▽ **Hendrickson H**

**Date** 1974 **Origin** USA

**Engine** Cummings turbo diesel

**Payload** Variable

Hendrickson trucks were favourites in the construction industry, as many were built with all-wheel drive. This model was a "prime mover", meaning a heavy haulage truck, often used to transport excavators and large construction machinery.

**Vertical** chrome exhaust pipe

**Optional air suspension** on rear axles

△ **Western Star 4964**

**Date** 1975 **Origin** USA

**Engine** Detroit diesel engine

**Payload** Variable

Western Star was formed in 1967 as a new truck brand from White. With all-wheel drive, Western Star rigs found a loyal following in Canada, especially for remote, off-road jobs.

**Fifth-wheel** trailer coupling

◁ **Dodge 950 Bighorn**

**Date** 1975 **Origin** USA

**Engine** 400 Cummins engine and various Detroit diesels

**Payload** Variable

One of the last big trucks built by Dodge, only 261 of these Bighorns rigs were made between 1973 and 1975. Only 60 of those built are known to exist today.

# Smaller Trucks from the 1970s

Economic disruption and fuel-supply concerns led to greater demand for lighter and more frugal commercial vehicles around the world. Haulage operators increasingly began to use smaller trucks, since vehicles under 7.5 tonnes (7.4 tons) did not necessitate a heavy goods licence in most countries. Away from heavy haulage, miniature trucks found multiple uses in local communities, and off-road vehicles were adapted to road use, as infrastructure improved. By this decade, vans and van-based pickups mostly sported the cab-over design, although some models still featured a short bonnet.

### ▽ Mercedes-Benz LP 608

**Date** 1975 **Origin** Germany

**Engine** 4-cylinder diesel, 79 hp

**Payload** 3.6 tons (3.5 tons)

Introduced in 1965, the LP range helped to establish the square cab-over design among light to medium road-haulage trucks in Europe. The LPs used interior fittings from some of Mercedes' car models.

Rear light bar

### ◁ Bajaj Pickup

**Date** 1971 **Origin** India

**Engine** 1-cylinder, 11 hp

**Payload** 175 kg (386 lb)

Powered by a small Vespa scooter engine, this small truck is still to be seen carrying out everyday tasks in India's cities and rural areas. Similar to the Ape (see opposite page), it was built under licence from Piaggio by Bajaj Auto.

### △ Mercedes-Benz Unimog 406

**Date** 1974 **Origin** Germany

**Engine** 6-cylinder diesel, 84 hp

**Payload** 2.7 tonnes (2.7 tons)

The Unimog was first made in 1948 by Boehringer. Daimler-Benz took over production in 1951. The 406 is a rugged off-road 4x4 truck with high ground clearance that makes it ideal for use in agriculture, forestry, construction, and the military.

**Tipper bed** is hydraulically operated

---

TECHNOLOGY

## Faun Kraka

The German Faun Kraka was a small, portable, utility vehicle that could fold to two-thirds of its normal length when not in use. Developed in 1962 originally for agriculture and forestry, the Faun Kraka's size enabled it to pass between trees and move more easily through undergrowth than larger vehicles. The German Army realized that it might have value in a military environment, particularly in airborne assault operations, when it could be deployed by parachute. After testing and modification, it was adopted by the army, serving from 1974 until the late 1990s. In its military incarnation, it was powered by a BMW 700 engine with 26 hp, and had a payload capacity of 750 kg (1,653 lb). Fully extended, the Faun Kraka was about 2.8 m (9 ft) long, but when folded its length reduced to around 1.8 m (6 ft).

Stage 1 – Fully extended

Stage 2

Stage 3

Stage 4 – Fully retracted

**Short rear corner bumpers**

### ▷ Fiat 241

**Date** 1974 **Origin** Italy

**Engine** 4-cylinder diesel, 47 bhp

**Payload** 1.4 tonnes (1.4 tons)

The rear-wheel-drive 241 was Fiat's attempt to capture a slice of the lucrative light-van market. Made from 1965 to 1974, it appeared as a direct contender to the Volkswagen Type 1 – the Fiat had a higher payload and was also a little faster. This model featured a 24-volt electrical system.

### ▷ Steyr-Daimler-Puch Pinzgaue

**Date** 1975  **Origin** Austria

**Engine** 4-cylinder petrol, 85 hp

**Payload** 1 tonne (1 ton)

Liked by European armies for its simplicity and cargo space, this 4×4 (also available as a 6×6) carried 10 passengers and could reach 109 km/h (68 mph). It had a backbone-tube chassis and coil sprung axles.

### △ Bedford CF

**Date** 1975  **Origin** UK

**Engine** 4-cylinder petrol, 79 hp

**Payload** 965 kg (2,127 lb)

The CF was intended as a direct rival to Ford's Transit. Exported around the world, both in van form and as a pickup, the CF was offered with petrol or diesel engines, and with 4- or 6-cylinders.

### ▽ Piaggio Ape

**Date** 1977  **Origin** Italy

**Engine** 1-cylinder, 11 hp

**Payload** 175 kg (386 lb)

Based on Piaggio's Vespa scooter, the Ape used handlebar steering. When it launched in the 1940s, the cab was open, but later was fully enclosed as in this model. The Ape is still in production today.

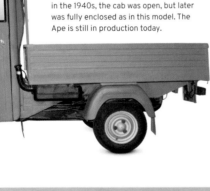

### ▷ Cushman Truckster

**Date** 1979  **Origin** USA

**Engine** 2-cylinder petrol, 32 hp

**Top speed** 200 kg (441 lb)

Scooter-manufacturer Cushman made this versatile three-wheeler. Available as a van or flatbed, the Truckster was used in a huge variety of roles, including ice cream van, mobile hot-dog stall, golf cart, and police vehicle. This one has a 3-speed gearbox and detachable doors.

**Side bars** stop vehicle from tipping over

**Battery location** under side panel

# Pickups from Around the World

As pickups became increasingly popular for leisure during the 1970s, the exact purpose of certain models at the time was ill-defined. Smaller pickups were becoming more popular and under the bonnet, engines varied greatly, from large, power-packing V8s to frugal little diesels. Some pickups were built purely for work, with their manufacturers prioritizing practicality and load-lugging ability, but those that doubled for both work and leisure were more popular. While some new models echoed heritage design, others looked to fresh horizons with bold new approaches. As a result, some pickups seemed afflicted by an identity crisis, their manufacturers not really knowing what they were for or who they were aimed at. This can be seen in the diverse range of pickups available at the time.

### △ Chevrolet El Camino SS

| | |
|---|---|
| **Date** 1972 | **Origin** USA |
| **Engine** V8, 325 hp or 450 hp | |
| **Payload** 300 kg (661 lb) | |

While Chevrolet had larger pickups in its range, the El Camino was without doubt the coolest. This SS (short for Super Sport) model was among the fastest pickups on the road at the time.

### ▷ Mercedes W115

| | |
|---|---|
| **Date** 1973 | **Origin** Germany |
| **Engine** 4-cylinder diesel, 60 hp | |
| **Payload** 300 kg (661 lb) | |

Fun, rugged, and dependable, this up-market pickup was heavily influenced by the stylish El Camino (above right). It was made for the Argentine market, but a ban on importing vehicles meant it had to be shipped in parts and assembled in the country.

**Cargo** protector

**Two-door, single cab** (a four-door, double-cab version was also produced)

**Ford name** on the bonnet

### ◁ Ford F-150

| | |
|---|---|
| **Date** 1975 | **Origin** USA |
| **Engine** 6-cylinder, 150 hp | |
| **Payload** 750 kg (1,653 lb) | |

Launched in 1975, the F-150 offered a combination of power and comfort that made it a runaway success. Ideal for work or as a commuter truck, it would become the bestselling US pickup.

### ▽ Land Rover Series III 109

| | |
|---|---|
| **Date** 1975 | **Origin** UK |
| **Engine** 6-cylinder petrol, 85 hp | |
| **Payload** 1.2 tonnes (1.2 tons) | |

Land Rover's Series III was launched in 1971. The heavy-duty, box-section chassis made this pickup incredibly rugged. A V8 pickup version was introduced in 1979, introducing a new flush-front design for these sturdy workhorses.

Cargo bed is 1.8 m
(6 ft) long

▷ **Volkswagen Caddy Pickup**

**Date** 1978  **Origin** Germany

**Engine** 4-cylinder petrol, 78 hp

**Payload** 500 kg (1,102 lb)

Based on the Mk1 Golf, this light pickup
was launched in the US in 1978 as the
Rabbit. It was brought to Europe in 1982
and rebranded the Caddy. It came with
car-grade interior refinements and a
choice of petrol or diesel engines. An
optional reinforced fibreglass hard top
for the cargo bed turned it into a van.

▽ **Chevrolet K-10**

**Date** 1978  **Origin** USA

**Engine** V8, 160 hp

**Payload** 750 kg (1,653 lb)

The K-10 provided competition for Ford's
F-series. With its Chevrolet V8 engine, it
shared a performance bloodline with the
Corvette. Large numbers of K-10s entered
service with the US Army and Air Force.

Simple grille of
stamped sheet metal

△ **Austin Mini Pickup**

**Date** 1979  **Origin** UK

**Engine** 4-cylinder petrol, 55 hp

**Payload** 250 kg (551 lb)

The iconic Mini was launched as a pickup
(based on the Mini's van platform) in 1961.
Compact, practical, and charming, it
remained in production until 1983, with few
changes. More than 54,000 were built.

▽ **Peugeot 504 Pickup**

**Date** 1979  **Origin** France

**Engine** 4-cylinder petrol, 62 hp

**Payload** 1.2 tonnes (1.2 tons)

A separate chassis enabled the 504 to be
offered with a variety of body styles, from
open pickup, as shown here, to box-body
van. One 504 pickup was successfully
raced as an off-road Group B rally car.

All-steel body

A Benz Lieferungswagen
from 1896

# Key Manufacturers
# The Mercedes-Benz Story

Daimler and Benz were pioneering rivals in the German truck industry. After pooling their resources in 1926, they went on to dominate, first the European, and then world markets with their high-quality engineering and enormous range. Today, the company is still breaking new technological ground with its trucks.

**THE TWO FAMOUS CREATORS** of the internal combustion engine independently founded the commercial vehicle sector in 1896. Gottlieb Daimler's Daimler-Motoren-Gesellschaft produced the first motorized truck that year, the Phoenix. Its 1-litre, 4-hp engine sent power to iron-clad wheels. Karl Benz, meanwhile, built the first van for local deliveries. With a payload of 300 kg (660 lb), the van had a single-cylinder, 5-hp engine and a three-speed transmission.

Surprisingly, they were both delivered to customers outside Germany, the truck to London and the van to a Parisian department store. For the first few weeks of its service in the UK, the truck was restricted to 4 mph (6 km/h) and had to be preceded by a man carrying a red flag. When these restrictions were lifted in November 1896, the era of motorized goods delivery took off.

By 1900, Benz & Cie had developed a range of its own trucks, the heaviest being a 5-tonne (4.9-ton) model with a 14-hp, two-cylinder engine, and a more robust chain drive, rather than Daimler's belt-driven transmission. The move to front-mounted engines and rubber tyres coincided with

**Karl Benz (1844–1929) is known as "the father of the automobile industry".**

Daimler's takeover of car manufacturer, Süddeutsche Automobilfabrik (SAG), with all truckmaking then centralized at the SAG plant at Gaggenau. The world's oldest truck factory, it still operates today.

Thanks to government subsidies, German truckmakers had a huge boost during World War I, and Daimler introduced the Marienfelde.

**Workers standing atop a Lo 2000 bus, c.1932**
This bus was made from the Lo 2000 platform truck. The world's first light-duty truck with a diesel engine, it paved the way for diesel to become the dominant fuel in the sector.

**Lastwagen**

**L337**

**1626L**

**Actros L**

| | |
|---|---|
| **1896** | Daimler introduces the world's first truck |
| **1896** | Benz launches the first delivery van, which still has similarities to a horse-drawn vehicle |
| **1899** | Benz launches its competitor truck to rival Daimler |
| **1926** | Daimler and Benz merge to form Daimler-Benz |
| **1939** | During WWII, company truck factories are heavily attacked by Allied bombs |
| **1949** | The L3250 starts postwar production |

| | |
|---|---|
| **1951** | Acquisition of agricultural vehicle manufacturer Unimog |
| **1958** | Mercedes-Benz trucks now assembled in 24 factories worldwide |
| **1963** | Introduction of LP 1620 with cab-over design rather than bonneted truck |
| **1970** | Takeover of rival commercial vehicle company Hanomag |
| **1973** | Tilt cabs introduced on the "New Generation" range |
| **1981** | US truck manufacturer Freightliner is taken over by Daimler-Benz |

| | |
|---|---|
| **1981** | The anti-lock braking system developed by Daimler-Benz is introduced to trucks |
| **1983** | A new range of rigid eight-wheelers is developed for the construction industry |
| **1991** | Development of experimental LEV – Low Emission Vehicle – to reduce car pollution problem |
| **1994** | Launch of NECAR 1, Europe's first working fuel-cell prototype van |
| **1995** | Debut of the popular Sprinter van range to replace the T1 Transporter |

| | |
|---|---|
| **1997** | Daimler-Benz-owned Freightliner takes over Ford's heavy truck-making operations in the US |
| **2000** | Daimler-Benz acquires US heavy truckmaker Western Star and power unit manufacturer Detroit Diesel |
| **2018** | Mercedes-Benz becomes first truck manufacturer to replace traditional wing mirrors with internal cameras |
| **2021** | Daimler Truck now standalone company |
| **2023** | Premiere of battery-electric eActros 600 with a range of 500 km (300 miles) |

**Gottlieb Daimler (1834–1900) introduced the world's first truck.**

Between 1914 and 1918, Daimler built 3,000 of this new four-cylinder, 3-tonne (3-ton) truck that was capable of 30 km/h (19 mph).

In 1926, Daimler and Benz merged as Daimler-Benz, and all cars and commercial vehicles were sold under the Mercedes-Benz name (Mercedes having been Daimler's car brand since 1901). Benz had introduced diesel engines in 1923, and once a Bosch injection pump had been added in 1927, the gutsy Mercedes-Benz Lo 2000 light truck and heavy-duty L5 became big sellers.

After World War II, demand for trucks – essential for rebuilding – was brisk. This demand was met by the 1949 diesel L3250, which was powerful, thrifty, and versatile, with its several wheelbase choices, plus a heater as standard. At the

same time, the 170D car-derived van and pickup were much in demand, too. In 1951, Daimler-Benz acquired the Unimog project, adding these four-wheel drive, high-riding hybrid truck/tractor multi-purpose vehicles to its line-up. The chassis formed part of the flexible suspension. Unimog vehicles are still built today.

In 1956, Mercedes-Benz launched the L319 panel van (the later L406 van of 1967 saw wider European success, as did the 1977 TN). Then, in 1958, it produced the LP333 platform truck, with its cab-over-engine layout, it gained the nickname "Millipede" for its two steered front axles. At 16–32 tonnes (16–31.5 tons), it was built for maximum payload and was comfortable for the driver.

Conventional and super-tough Mercedes-Benz short-nosed trucks were in growing demand the world over and were made right up to the 1990s. However, the LP1620 of 1963 became a modern milestone with its

cubic cab offering space and large windows, thanks to minimal engine intrusion. The cab was systematically adapted to a wide range of models and helped Mercedes-Benz become Europe's biggest manufacturer, with its annual output tripling between 1960 and 1970.

In 1973, a new basic modular truck design led to 76 derivatives in the space of a year. Updated with heavy-duty axles, they were capable of driving over the most extreme terrains. Technology was also applied to safety in the 1980s, with electronic anti-lock brakes in 1981 greatly cutting the risks of jack-knifing and trailer disconnection. The system was soon mandatory across the industry, and further safety improvements followed with the aerodynamic Actros series – introduced in 1996, the company's centenary year. These included anti-skid control, lane-keeping assist, and active sideguard-assist.

**Mercedes-Benz advertisement, c.1949**
With trucks in high demand after WWII, Mercedes-Benz launched a series of diesel models, beginning with the L3250 in 1949, followed by the L3500 and L4500.

The Actros series was joined by the broad Atego light truck range two years later, hauling 6.5–16 tonnes (6.4–15.7 tons). Cockpit cameras replaced cab mirrors in the Actros in 2018, and production began for its fully electric eActros model three years later.

Mercedes-Benz also developed a range of vans alongside its trucks. In 1995, its Sprinter range became an instant global hit, later joined by the more compact 1996 Vito built in Spain, and then the ultra-small Citan (a rebadged Renault Kangoo).

Mercedes-Benz cars and trucks were finally separated into separate entities in October 2021 when Daimler Truck was spun off as a separate company, although carmaker Daimler AG retained a 35 per cent stake.

**Merging identities**
When Daimler-Benz acquired the semi manufacturer Freightliner in 1981, it was already a long-established and iconic truck brand in the US.

# Mega Earth Movers

The mining and quarrying sectors require specialist trucks to move huge amounts of excavated earth and stone. They must be large and powerful enough – often massively so – in order to carry their huge loads, yet must also be capable of operating reliably in an extreme environment.

## HEAVY LOAD

Since the 1950s, trucks designed for mining and quarrying operations have continuously evolved to be as efficient – and stable – as possible. By the 1970s, trucks of a scale such as this were becoming commonplace – vehicles grew in size as technology advanced. These trucks, called front-end loaders, move heavy loads with enormous buckets mounted on the front, which means that they must have sufficient weight at the back to counterbalance it. They also need extra grip for challenging terrains, so the oversize tyres are fitted with tracks. Drivers must be highly skilled to negotiate the tracks without losing any of their load or rolling their vehicle off the side. They also need a head for heights just to access the cab – some are as high as eight storeys from the ground.

**A front-end loader mining truck** at a coal mine in the late 1970s. It is also called a "wheel loader" as it uses wheels, instead of tracks, for enhanced speed.

# Trucks Hurtle into Hollywood

Trucks have long graced the screen, from Laurel and Hardy films in the 1920s through *Terminator 2*'s (1984) memorable truck sequence to the present-day *The Fast and the Furious* series (2001–). Menacing, exciting, aspirational, and funny, both in supporting and in starring roles, trucks have produced some iconic movie and TV moments.

The first half of the 20th century saw several films in which trucks played a key role, including *They Drive by Night* (1940), with a pre-*Casablanca* Humphrey Bogart, a truck-based Soviet musical comedy called *Happy Flight* (1949), and the French thriller *The Wages of Fear* (1953). In 1957's *Hell Drivers*, ex-convict Tom Yately (Stanley Baker) drives a Dodge 100 "Kew" truck for a crooked foreman who pushes drivers too far. Forcing them to deliver extra loads at breakneck speed over deadly routes, the foreman pockets the extra cash himself. After a driver's death triggers a showdown, another Dodge, with foreman inside, plunges into a quarry.

In terms of vehicles, the car became the star first, in films such as 1968's Steve McQueen hit *Bullitt*, with its groundbreaking car chase, the hot-rod delivery driver in a Dodge

Challenger in *Vanishing Point* (1971), and *American Graffiti* (1973) examining the cruising culture of early 1960s America. It is not at all surprising that trucks then muscled their way onto centre stage.

## Trucks in starring roles

Taking road chases to a different level, Stephen Spielberg's *Duel* (1971) introduced a menacing and relentless

**Duel, 1971**
In Spielberg's *Duel*, salesman David Mann (Dennis Weaver) is terrorized during a business trip through California by the driver of a semi-truck. The driver is virtually unseen during the entire film, so the effect is that of being hunted down by the truck itself.

18-wheeled antagonist. The film won two Emmy nominations for sound and cinematography, but the truck was clearly the star of the show. The rusty, weather-beaten 1955 Peterbilt 281 was equipped with a 260-hp, turbo-diesel engine that could reach 129 km/h (80 mph) – perfect for a high-speed car chase.

In 1977's beer-smuggling comedy, *Smokey and the Bandit*, Burt Reynolds and his famous Pontiac Trans Am were joined by a 1974 Kenworth W900A, one of the truly iconic American big rigs. A year later, trucks were the headline act of director Sam Peckinpah's film *Convoy*, in which another legendary American truck, a Mack RS700L, was accompanied by a cast of Peterbilts and Kenworths. In reality, several Macks stood in for the main vehicle, with some of them rigged with cameras and special effects apparatus (all part of the Hollywood magic).

**Convoy, 1978**
In this film, driver Martin Penwald (call sign "Rubber-Duck"), played by Kris Kristofferson, inadvertently becomes the leader of a trucker revolt, heading a convoy of haulage vehicles through Arizona, New Mexico, and Texas. Here, a police car gets sandwiched between the rigs of truckers "Pig Pen" and "Spider Mike".

*Convoy* gave trucks some cinematic cool, and although the citizens' band (CB) radio craze was already building, the movie gave it a phenomenal boost. Before the age of mobile phones, CB radio was a valued means of communication, with drivers adopting their own call signs, or "handles". It became a lifeline for truck drivers, not only giving them some welcome social interaction on long trips, but also enabling them to give each other an invaluable "heads up" of what lay ahead on the road.

Big semi-trucks have also been the focus of documentary films, with trucking coal across the Gobi Desert to China being the subject of 2022's *Lady of the Gobi*. The film, which shows the

THE RANK ORGANISATION
**STANLEY BAKER · HERBERT LOM · PEGGY CUMMINS**

# HELL DRIVERS

**VISTAVISION** · **PATRICK McGOOHAN**

also starring

Screenplay by JOHN KRUSE and C. RAKER ENDFIELD · Produced by S. BENJAMIN FISZ · Directed by C. RAKER ENDFIELD

**Hell Drivers, 1957**
The trucks used in *Hell Drivers* were built at the Dodge factory in Kew, London. They were nicknamed "parrot noses", due to the shape of their grille and bonnet.

# "Boy, these **lonely long highways** sure grind the souls of us **cowboys.**"

BOBBY "LOVE MACHINE" "PIG PEN", *CONVOY*, 1978

dangers of the trucking life, follows driver Maikhuu along Mongolia's coal highway. She encounters land mines, accidents, and pollution as she heads to the Chinese border in search of a better life for her family.

## Small screen action

In addition to the silver screen, trucks also became stars of the small screen, particulary during the 1980s, when they featured in a host of popular action TV shows for children. One of the most memorable was a modified Peterbilt 352 Pacemaker that provided a mobile refuge for Kit, the talking car, in *Knight Rider* (1982–86).

Also aimed at younger viewers, *The A-team*'s battle bus, a GMC Vandura, first hit TV screens in 1983, and stuntman Colt Seavers (Lee Majors) trashed numerous GMC K2500 Sierra pickups in the jump scenes of *The Fall Guy* (1981–86).

More recently, the trend for "extreme" trucking reality TV shows aimed at adult viewers began with *Ice Road Truckers* (2007–17), which shadowed drivers taking trucks over frozen lakes and rivers in Canada and Alaska. Continuing the theme were *Outback Truckers* (2012–), *Trucking Hell* (2018–), and French series *Les Reines de la Route* (2021–).

### Character-building Trucks

Some trucks have found screen fame in animated form. The *Scooby-Doo* Mystery Machine, which first rolled into view in 1969, has changed its form over the years. It was never meant to be an accurate rendition of a specific vehicle, but 2002's *Scooby-Doo* movie used a Bedford CF. Perhaps the most famous truck character is Optimus Prime from the *Transformers* franchise. Originally, Prime was disguised as a Freightliner FL86, later, a Kenworth K100 Aerodyne. For the first movie in 2007, he became a Peterbilt 379, then shifted again into a Western Star 5700XE in *Transformers: The Last Knight* (2017).

**Fans argue about the true identity** of the *Scooby-Doo* Mystery Machine. Influences from Chevrolet, Dodge, Ford, and VW can be seen in its various incarnations across the decades.

# Unimog 406 U900

The Unimog is Mercedes-Benz's bestselling all-terrain vehicle. The story began in 1946 when engineers Albert Friedrich, Heinrich Rößler, and Hans Zabel realized the demand for a small, versatile, road-going tractor that could be adapted to a variety of agricultural and forestry work. Production began in 1948, but original maker Boehringer struggled to keep up with demand and handed over manufacture to Daimler-Benz from 1951.

**IN ESSENCE, ALL UNIMOG MODELS** are extreme off-roaders, with four-wheel drive and high ground clearance. They can go anywhere and do anything, as demonstrated by their ability to power an extensive range of machinery from all sides – not just the rear, as most conventional tractors with a single power take-off feed from the engine do. The 406 model shown here began production in 1963, which many regard as the benchmark Unimog. These versatile vehicles have been in the service of many armed forces around the world and are widely used in land-based industries.

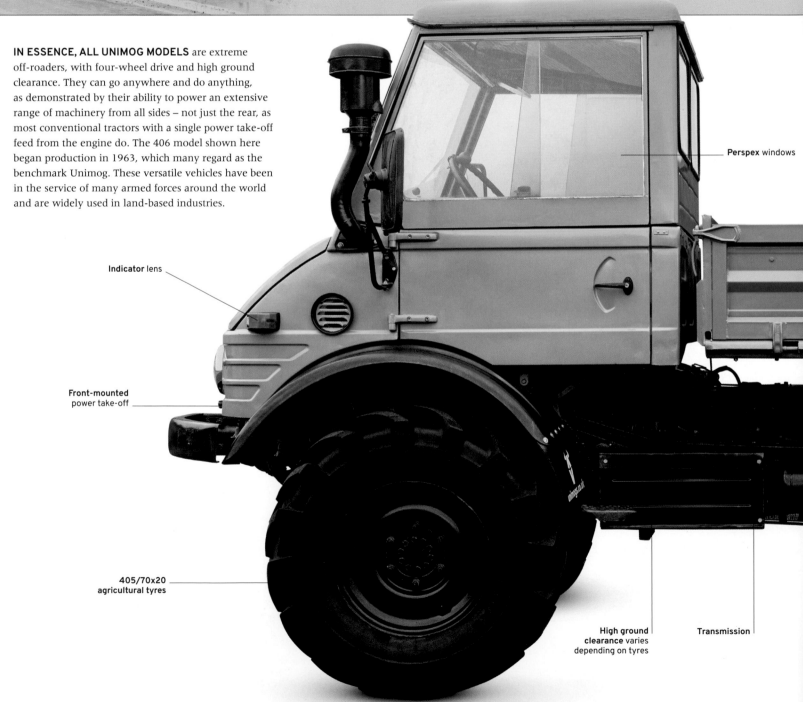

**Perspex** windows

**Indicator** lens

**Front-mounted** power take-off

**405/70x20** agricultural tyres

**High ground clearance** varies depending on tyres

**Transmission**

| SPECIFICATIONS | |
| --- | --- |
| Model | Mercedes-Benz Unimog 406 U900 |
| Origin | Germany |
| Assembly | Mercedes-Benz-Werk Gaggenau |
| Production run | 1963–89 |
| Weight | 3.3 tonnes (3.2 tons) |
| Load capacity | 2.7 tonnes (2.7 tons) |
| Engine | 6-cylinder diesel, 84 hp |
| Transmission | 6-speed manual/20-speed cascade |
| Maximum speed | 71 km/h (44 mph) |

**Universal name**
The name "Unimog" was created by engineer Hans Zabel and was an abbreviation of the title given to the prototype concept: *Universal-Motor-Gerät für Landwirtschaft* ("Universally applicable motorized machine for agriculture").

**External** bonnet release hole

FRONT VIEW

**Drop-down** side panels

**Cargo bed**

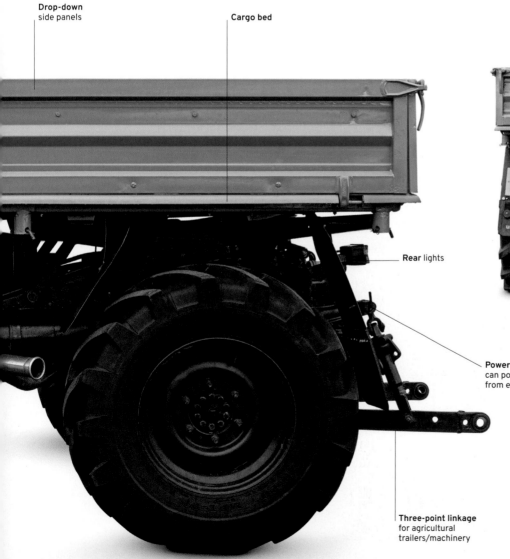

**Rear** lights

REAR VIEW

**Power take-off output** can power machinery from engine

**Three-point linkage** for agricultural trailers/machinery

**Versatile vehicle**
The versatility of the Unimog means it can even be fitted with rail wheels, so it can be used on railway lines for maintenance work or as shunters. On farms, Unimogs are used to carry out the work of conventional tractors – ploughing and cultivating.

## THE EXTERIOR

This example has a high-top cab, with a higher roofline than standard, but many 406s and subsequent models had canvas convertible roofs that allowed for better visibility. It is also fitted with a snorkel to allow the engine to draw in air while driving through deep water. With its independent coil-sprung suspension, this model also has portal axles positioned above the wheel centre that brings the drivetrain higher into the chassis. This allows for greater ground clearance and flexibility – making this vehicle almost unstoppable.

**1.** Bonnet badge  **2.** Air intake  **3.** Fuel-filler neck
**4.** Front power take-off shaft  **5.** Gearbox and transmission propshaft with power take-off lever  **6.** Loading-bed panel catch  **7.** Updated rear-light units  **8.** Coil spring with helper springs  **9.** Rear power take-off drivetrain

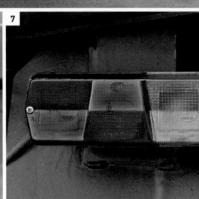

## THE ENGINE

The 406 uses a Mercedes diesel truck engine that generates 84 hp in its standard form. Simple and reliable, the direct-injected OM352 engine has a capacity of 5,675 cc normally aspirated, although some have been converted to turbocharged engines. The gearbox has a separate 20 "cascade" selector that enables the Unimog to operate at a range of extremely slow crawling speeds.

**10.** Engine bay

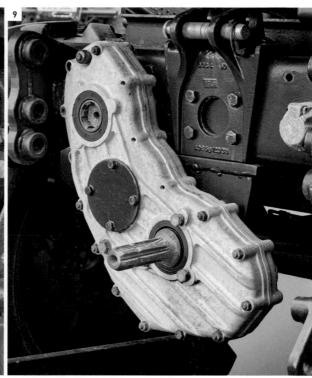

## THE INTERIOR

There is no carpet and little in the way of plastic trim inside this 406. Everything about it was designed for withstanding the elements in an open cab: most of the electrics are well sealed in watertight spaces and the doors are easily removable. A hand throttle on the dashboard is used to set the revs when running machinery, idling, and when a constant slow speed is required. The transmission levers consist of the standard six-speed gearbox selector, a forward/reverse engagement lever (allowing all gears to be used in reverse), and other controls such as a power take-off lever.

**11.** Driving area with central engine cover
**12.** Engine starting controls   **13.** Transmission and range changing levers   **14.** Cab door spring
**15.** Cab interior door release

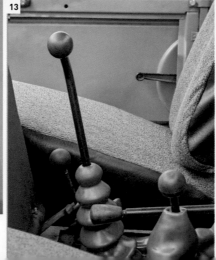

# Child's Play

The first toy vehicles appeared soon after the invention of the car, with tinplate cars being produced in the early 1900s. Diecasting with zinc alloys later reduced costs, and eventually new plastic technology enabled authentic-looking models to be made in seconds. Realizing the allure of huge machines to children, toy manufacturers inevitably turned their attention to trucks of all kinds.

## THE TONKA STORY

Tonka is one of the best-known toy truck makers. It launched its first models – a steam shovel and a crane – in the US in 1947. Their runaway success led Tonka to extend its range to a wide variety of truck types and construction vehicles, including a pickup (1955), a Jeep (1962), and the iconic, bright yellow Mighty Dump Truck (1965). Made of pressed automotive steel with real rubber tyres, Tonka toys earned a reputation for both realism and durability, able to withstand whatever rough play children put them through. The line-up has changed over the years, with models being offered in plastic as well as metal. Today, vintage Tonka trucks are collector's items that can command big prices at auction.

**Tonka truck parade:** even though Tonka toys are made to last, rust is a common problem, but with care they can be restored to their former glory.

# Japanese Light Trucks

Miniature vans and pickups are very much a part of modern Japanese culture, much-used by businesses, communities, and individuals. In the 1970s, their size and design was heavily influenced by a classification called "Kei Truck", which specified strict engine capacities to define light vehicles for taxation purposes. These rules applied only to domestic vehicles, and many export models grew larger during this period, with the Toyota Land Cruiser and Nissan Patrol among those achieving overseas success. Smaller trucks such as the Honda Acty and Suzuki Carry were also introduced, with fuel-efficient 4-cylinder engines later replacing older, noisy 2-stroke units. Japanese pickups of the 1970s gained a global reputation for reliability.

Reverse-opening doors

### △ Subaru Sambar Pickup

**Date** 1970 **Origin** Japan

**Engine** 2-cylinder, 2-stroke, 25 hp

**Payload** 420 kg (926 lb)

Measuring just 2.99 m (9.8 ft) in length, this charming little second-generation Sambar pickup had a tiny rear-mounted, air-cooled 356 cc engine. The Sambar was one of the first small Japanese pickups designed around the Kei Class specifications to meet taxation rules.

### ▽ Subaru Sambar Minivan

**Date** 1973 **Origin** Japan

**Engine** 2-cylinder, 2-stroke, 28 hp

**Payload** 420 kg (926 lb)

In 1973, the Sambar received a facelift and a water-cooled engine in its third-generation reboot. Removable seats enabled this versatile minivan version to be quickly transformed from a passenger vehicle into a load-carrier.

Sliding
side doors

### ▽ Honda Vamos

**Date** 1970 **Origin** Japan

**Engine** 2-cylinder air-cooled, 30 hp

**Payload** 220 kg (485 lb)

The small, open Vamos was a recreational vehicle that could carry a small payload in the rear, making it ideal for beach trips. Other similar designs on the market included the VW Thing and the Mini Moke. Only 2,500 Vamos vehicles were made.

Side rails for
passengers to hold

Cargo rails

Steel
bodywork

Engine bay cover

### ▷ Datsun 620 Pickup

**Date** 1976 **Origin** Japan

**Engine** 4-cylinder petrol, 52 hp

**Payload** 600 kg (1,323 lb)

With global sales of Japanese vehicles soaring in the 1970s, the 620 found success in North American and European markets. This example has been modified with updated wheels and lowered suspension.

Traditional-style
wing mirror

Modern
alloy wheels

◁ **Toyota Land Cruiser FJ45**

Date 1976  Origin Japan

Engine 6-cylinder, 133 bhp

Payload 1 tonne (1 ton)

The FJ series Land Cruiser was inspired by versatile 4×4 vehicles, such as the Land Rover and Jeep. The rear body tub on this FJ45 long-wheelbase pickup could be removed from the chassis.

**Fire hose** supplied by onboard water tank

**Fire siren**

**One-piece windscreen**

△ **Nissan Patrol Fire Truck**

Date 1978  Origin Japan

Engine 6-cylinder, 128 hp

Payload 1 tonne (1 ton)

Nissan's answer to Toyota's Land Cruiser, the Patrol became a global success. With a powerful engine and good manoeuvrability, this small fire truck version could be quickly deployed at a blaze.

**Car-derived front** and engine

BRP 349S

△ **Toyota Hilux N20**

Date 1978  Origin Japan

Engine 4-cylinder petrol, 81 hp

Payload 635 kg (1,400 lb)

The first Hilux appeared in 1968. This second-generation, car-derived N20 offered automatic or manual transmission. A major redesign would soon turn the Hilux into its own distinctive brand.

**Regulation lights:** despite resembling a buggy, the Vamos was fully road legal

**Fixed side panels**

**Engine snorkel** delivers air to engine when driving through water

**Drop-down tailgate**

▷ **Toyota Land Cruiser HJ45**

Date 1979  Origin Japan

Engine 6-cylinder, 133 bhp

Payload 1 tonne (1 ton)

Easy to modify, seemingly indestructible, and with great towing ability, this crew-cab version of the FJ was ideal for use in remote locations, and it proved especially popular in Australia. Station wagon variants were also available.

# 1980–1999
# TURBO POWER

# TURBO POWER

**The era of composite materials**, microchips, and speed limiters, the 1980s and 90s brought advanced technology into trucks of all sizes, with both positive and not-so-positive outcomes. Pickup trucks surged in popularity during this time, perhaps inspired by blockbuster films such as *Back to the Future* (1985) that made a star of Toyota's go-anywhere Hilux model, allowing it to steal sales from US competitors. Customization was also a craze, as owners sought to personalize their pickups. Seeing an opportunity, manufacturers then began offering high-performance models, such as the V10-powered Dodge Ram.

In Italy, IVECO began a joint venture with Ford to increase output of their European trucks, while the diminutive Bedford Rascal microvan – a collaboration between GM and Suzuki – brought miniature movers to the masses.

With complex electronics more readily available, leading manufacturers introduced new semi-automatic gearboxes in a bid to poach sales from their competitors. The early days of these new transmissions marked the dawn of an era of almost effortless truck driving, while purists still preferred the traditional manual stick for reliability and control.

At the end of the last century, trucks could still be maintained easily on the road by a single operator. After this point, electronic wizardry took hold, which made it challenging for drivers to carry out repairs themselves. However, this new technology made trucks much safer and more fuel efficient.

△ **Truck carrying timber, 1983**
A heavy-duty truck collects a load of timber from along a dirt logging road. This all-terrain model is a first-generation truck put out by Russian manufacturer KamAZ.

## Key events

▷ **1980** Scania 2 series of commercial trucks is introduced in Europe.

△ **Scania T142H, 1980**
Part of Scania's GPRT or 2 series, the T142H was a heavy-duty conventional cab truck with a 14.2-litre engine. The "T" in its model name stood for "Torpedo".

▷ **1981** Dodge Ram launches – the iconic model becomes a brand and quickly gains a loyal following.

▷ **1982** DAF truck wins the Paris-Dakar Rally, which tests all-terrain vehicles with a 10,000-km (6,200-mile) race across the Sahara Desert.

▷ **1984** Land Rover Defender (initially released as the Land Rover 90) launches with coil-sprung suspension.

▷ **1986** Tata 407 light commercial vehicle enters production in India.

▷ **1986** Tachographs to regulate driver hours become mandatory in Europe for trucks over 3.5 tonnes (3.4 tons).

▷ **1990** The Renault Magnum truck is introduced with a radical and typically futuristic Renault design.

▷ **1994** Channel Tunnel opens between the UK and Europe. Trucks are served by a separate terminal and train.

▷ **1999** Peterbilt 387 launches. It is modern, sleek, and aerodynamic, yet with classic American styling.

"If the **1980s** had been the **decade of sophistication** for the truck, the **1990s** was a decade devoted to **environmental considerations.**"

VOLVOTRUCKS.COM

◁ **An ornately decorated Bedford truck** transports goods in Pakistan, 1986.

# Transcontinental Models

In the 1980s and 90s, truck technology made rapid advances, and trucks became much more like the models seen on roads today. As pan-European haulage became increasingly common, sleeper cabs and enhanced driver comfort were focuses for manufacturers, as were more powerful engines to haul heavier loads. Environmental issues were also being taken seriously, and engine emissions began to be regulated by the European Union. This era saw large numbers of trucks sold across Europe, some of which are now iconic for their innovation, comfort, and performance. A number of European manufacturers, including Volvo, also turned their attention to boosting sales in the North American market.

**Bodywork**
streamlined for aerodynamic benefit

◁ **Leyland Roadtrain 20-32**

| | |
|---|---|
| **Date** 1987 | **Origin** UK |
| **Engine** 6-cylinder, 280 hp | |
| **Payload** 38 tonnes (37 tons) GVW | |

**Fuel tank**

◁ **Volvo F16**

| | |
|---|---|
| **Date** 1987 | **Origin** Sweden |
| **Engine** 6-cylinder, 460 hp | |
| **Payload** 38 tonnes (37 tons) | |

△ **DAF 3300 ATI SpaceCab**

| | |
|---|---|
| **Date** 1984 | **Origin** Netherlands |
| **Engine** 6-cylinder, 330 hp | |
| **Payload** 38 tonnes (37 tons) | |

In 1984, DAF introduced the SpaceCab on its existing 3300 flagship model. Ushering in a new era of spaciousness for long-haul cabs, it was the first in which a driver was able to stand fully upright.

The Roadtrain, Leyland's 1980s workhorse, was very popular as a fleet truck. This version features a high roof. The Roadtrain was renamed the Leyland DAF 80 series in 1991, after the merger of the two companies.

Volvo introduced its F16 tractor unit – so called as it had a 16-litre engine – in 1987. At the time, it was the most powerful truck on the market, boasting 460 hp, and it remained popular until it was replaced by the FH16 in 1993.

**Exhaust stack**

**Exhaust heat shield** made of perforated metal tubing

**Fifth-wheel** trailer coupling

**Tail light**

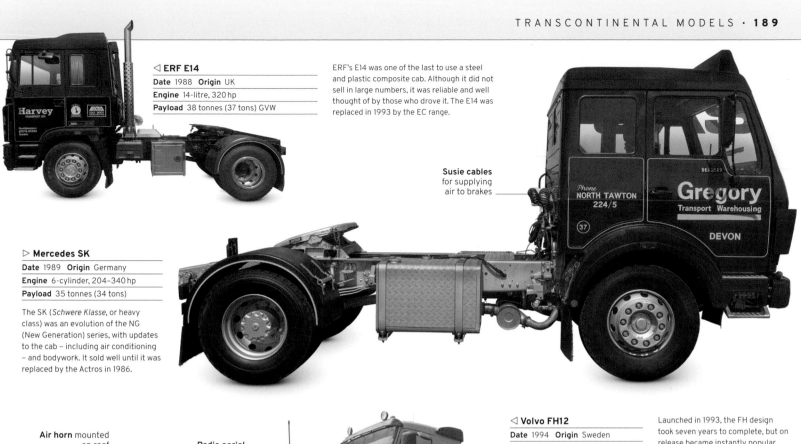

### ◁ ERF E14

| Date 1988 | Origin UK |
|---|---|
| **Engine** 14-litre, 320 hp | |
| **Payload** 38 tonnes (37 tons) GVW | |

ERF's E14 was one of the last to use a steel and plastic composite cab. Although it did not sell in large numbers, it was reliable and well thought of by those who drove it. The E14 was replaced in 1993 by the EC range.

**Susie cables** for supplying air to brakes

### ▷ Mercedes SK

| Date 1989 | Origin Germany |
|---|---|
| **Engine** 6-cylinder, 204–340 hp | |
| **Payload** 35 tonnes (34 tons) | |

The SK (*Schwere Klasse*, or heavy class) was an evolution of the NG (New Generation) series, with updates to the cab – including air conditioning – and bodywork. It sold well until it was replaced by the Actros in 1986.

**Air horn** mounted on roof

**Radio aerial**

**Hood grille**

### ◁ Volvo FH12

| Date 1994 | Origin Sweden |
|---|---|
| **Engine** 6-cylinder, 420 hp | |
| **Payload** 44–60 tonnes (43–59 tons) GVW | |

Launched in 1993, the FH design took seven years to complete, but on release became instantly popular because of its comfort and reliability. Global sales of the successful series exceed 1.4 million units.

### △ Renault Magnum

| Date 1995 | Origin France |
|---|---|
| **Engine** 6-cylinder, 374 hp | |
| **Payload** 44 tonnes (43 tons) | |

Launched in 1990 as the Renault AE (for AErodynamic), this truck won the International Truck of the Year title the following year. It was later renamed the Magnum, staying in production until 2013. It has become iconic for its high level of driver comfort and fully flat cab floor.

### ◁ Volvo VNL 640

| Date 1999 | Origin Sweden |
|---|---|
| **Engine** 6-cylinder, 350–420 hp | |
| **Payload** 40 tonnes (39 tons) GVW | |

Volvo's VNL (L standing for long bonnet – a style not common in Europe) was its first truck aimed at the North American market after it had taken over and then discontinued the WhiteGMC brand in 1997.

# Leyland Popemobile

Launched in 1981, the Leyland Constructor offered a modern take on truck styling, with smooth corners created in composite panels. Designed by British automotive specialist Ogle Design, well-known for its work with Aston Martin, the advanced T45 cab was said to have an aerodynamic profile 30 per cent more efficient than that of any truck on the market. In 1982, two Constructors were adapted for use by the Pope during a visit to the UK.

**THE LEYLAND CONSTRUCTOR** was a heavy-duty 6x4 and 8x4 truck manufactured from 1980 to 1993. It was intended to be used primarily as a tipper lorry, but such was its success both operationally and in terms of driver refinement that it also found other applications.

One such role was in the military, where it was used for rapid equipment deployment, helping to modernize the British army in the late 1980s and 90s. With ample power and a strong chassis, the truck could haul itself at speed over almost any terrain.

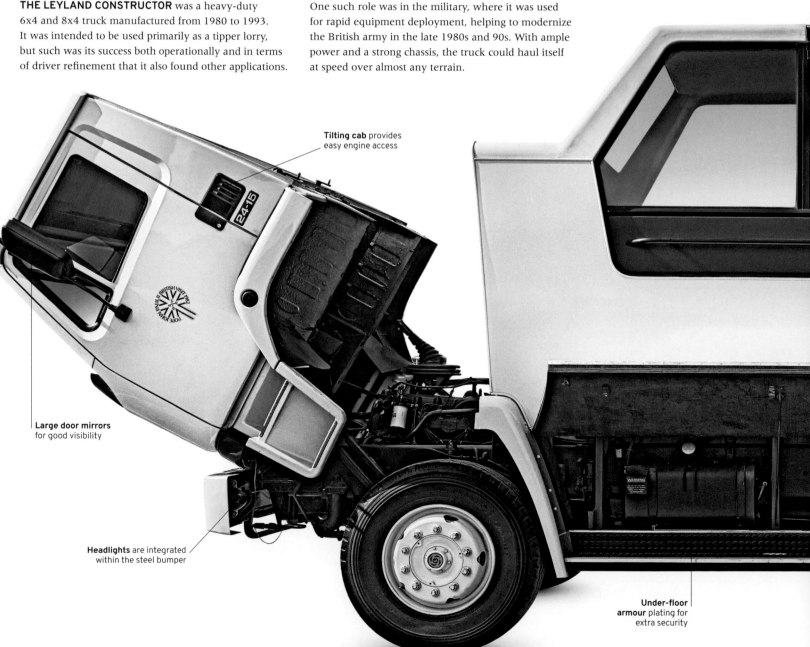

**Tilting cab** provides easy engine access

**Large door mirrors** for good visibility

**Headlights** are integrated within the steel bumper

**Under-floor armour** plating for extra security

| SPECIFICATIONS | |
|---|---|
| **Model** | 1982 Leyland Popemobile |
| **Origin** | UK |
| **Assembly** | Leyland, Lancashire |
| **Production run** | 2 |
| **Weight** | 22 tonnes (21.7 tons) |
| **Load capacity** | 9–27 tonnes (8.9–26.6 tons) |
| **Engine** | 6-cylinder Leyland TL11 turbo diesel, 181 hp |
| **Transmission** | 10-speed splitter |
| **Maximum speed** | 96 km/h (60 mph) |

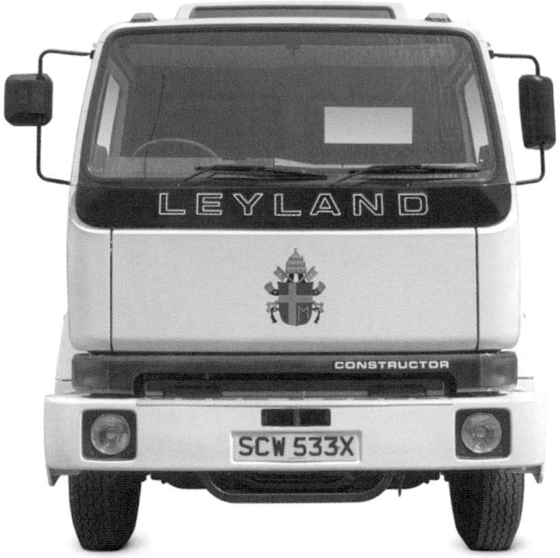
LEYLAND
CONSTRUCTOR
SCW 533X

FRONT VIEW

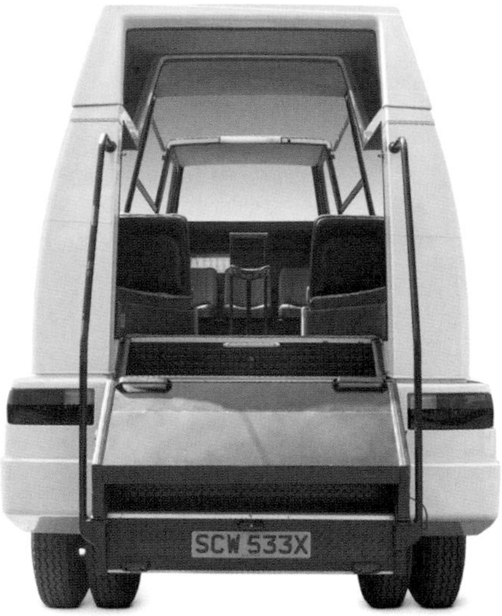
SCW 533X

REAR VIEW

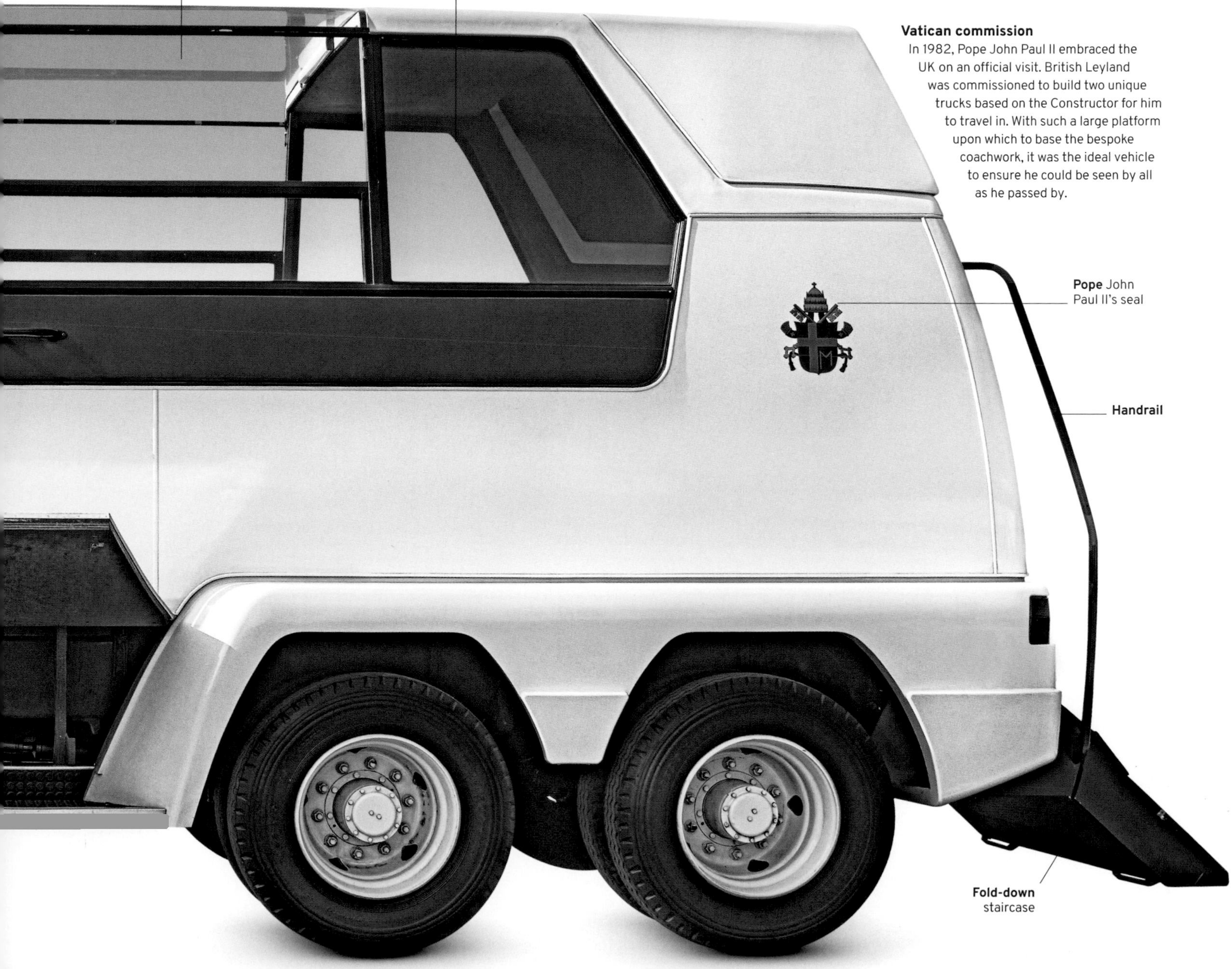
**Raised bodywork** provided a viewing platform for the Pope
**Bulletproof** glass windows
**Pope** John Paul II's seal
**Handrail**
**Fold-down** staircase

**Vatican commission**
In 1982, Pope John Paul II embraced the UK on an official visit. British Leyland was commissioned to build two unique trucks based on the Constructor for him to travel in. With such a large platform upon which to base the bespoke coachwork, it was the ideal vehicle to ensure he could be seen by all as he passed by.

## THE EXTERIOR

There are not many trucks with as brief a purpose as this one had. It was created for the Pope to be seen by the public on his UK tour in 1982, so the design had to be striking, while providing maximum security. A large amount of the truck's bodywork is bulletproof glass, raised in the centre to provide maximum standing headroom. Under-floor armour plating and emergency exits are also essentials. At the rear, a fold-out staircase looks more like something found on an aircraft than a truck. On either side of the bodywork, handrails and footrests allow bodyguards to travel with the vehicle on parade.

**1.** Pope John Paul II's seal  **2.** Door handle and unique engine badge  **3.** Leyland new cab lettering  **4.** The Pope's logo for his tour of Britain  **5.** Engine oil filler and dipstick  **6.** Front wheel with Leyland centre cap  **7.** Fold-down staircase at rear of vehicle

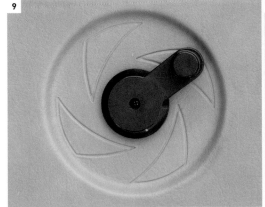

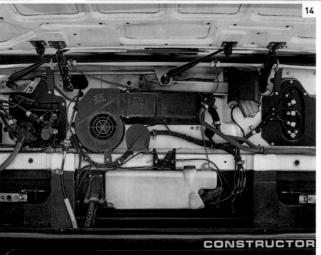

## THE INTERIOR

Featuring a fully adjustable steering rack, the cab also has generous sound-proofing. The fuse and relay box is housed on the dashboard to protect it from moisture and provide easy access for maintenance. To enable the Pope to both sit and stand while on the move, a seat with a handrail backrest was installed to ensure stability at slow parade speeds, with extra seating for staff at the rear.

**8.** Cab interior **9.** Cab window winder **10.** Panel fastening button **11.** Pope John Paul II's seat and support rail **12.** Staff seat

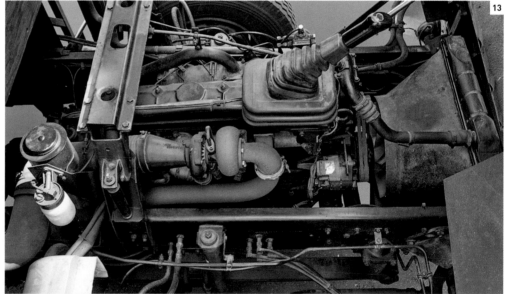

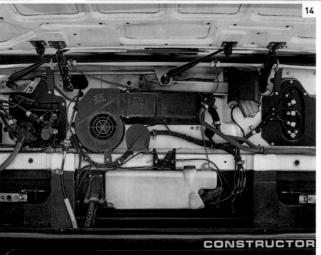

## THE ENGINE

Its tilting cab means the Leyland TL11 6-cylinder turbo diesel is easily accessible. While these engines proved to be both economical and reliable, they were eventually replaced by other options from Rolls-Royce and Gardner.

**13.** Engine and gearbox **14.** Cab auxiliary access panel

# 80s and 90s Pickups

In the 1980s and 1990s, pickup trucks grew in popularity across the world. US manufacturers, such as Ford and Chevrolet, which had led the way for years, produced some classic models during this era – some of which are now collectors' items. Japanese marques were also gaining popularity, with Datsun and Toyota producing models that became global successes. While many models were still being designed as mechanical beasts of burden, this era also saw the emergence of sports trucks, in which the focus was more on performance than utility.

### △ Ford F-250

**Date** 1980  **Origin** USA

**Engine** 6-cylinder, 120 hp

**Payload** 750 kg (1,653 lb)

Launched in 1980, the 250 was part of the seventh generation F-series – the range's first major redesign since 1965. Two-door, two-door extended, and four-door crew cab options were offered.

**Bonnet-mounted** wing mirror

### △ Datsun 720 Pickup

**Date** 1985  **Origin** Japan

**Engine** 4-cylinder, 103 hp

**Payload** 1 tonne (1 ton)

Datsun's pickup had been a fixture in its line-up for many years. It was popular in Japan and the US, where it was made at Nissan's new Tennessee plant. The 720 above is the flatbed long-body deluxe.

### ▽ Toyota Hilux

**Date** 1985  **Origin** Japan

**Engine** 4-cylinder, 94hp

**Payload** 955 kg (2,105 lb)

Hilux models have had different names in different territories. Known as the Toyota Pickup in the US, the 4x4 below was the fourth generation of the Hilux that launched in 1968.

### ▷ Datsun Sunny B120

**Date** 1987  **Origin** Japan

**Engine** 4-cylinder, 69 hp

**Payload** 500 kg (1,102 lb)

Lightweight and durable, the Sunny was a workhorse suitable for all manner of tasks. During its near-25 years of production, it changed very little. This model, the B120, was especially popular in South Africa.

## △ Chevrolet C3500 Silverado 3+3

**Date** 1989 **Origin** USA

**Engine** 6-cylinder, 143–230 hp

**Payload** 1 tonne (1 ton)

This Chevrolet Silverado 3+3 – so called because it came with two bench seats that could fit three people up front and three in the back – has been built as a recovery truck, showing the diversity of the model.

## ▽ Dodge Shelby Dakota

**Date** 1989 **Origin** USA

**Engine** 8-cylinder, 175 hp

**Payload** 567 kg (1,250 lb)

A limited edition, performance version of the Dakota Sport pickup, the rear-wheel drive Shelby was offered for 1989 only. Individually numbered dash plaques showed each truck's heritage.

**More powerful** engine than standard pickups

**All-black livery** sets the 454 SS apart from other models

## ⊲ Chevrolet 454 SS Pickup

**Date** 1990 **Origin** USA

**Engine** 8-cylinder, 230 hp

**Payload** 500 kg (1,102 lb)

The 454 SS Pickup was launched in 1990 to cash in on the sport-truck scene that had emerged in recent years. Its powerful engine, combined with black livery and red interior, made it a favourite from the start. It is now highly collectable.

**Front styling** resembles a Ford Sierra

## ▷ Ford P100

**Date** 1991 **Origin** Portugal

**Engine** 4-cylinder, 73 hp

**Payload** 1 tonne (1 ton)

This version of the long-running P100 was based on the Ford Sierra car and was renowned for its reliability and durability. Both diesel and petrol versions were offered. It has recently become popular in modifying and drifting circles.

**Higher ride height** than the Sierra, due to larger wheels and multi-leaf spring suspension at the rear

## ▽ Mercedes 290 GD

**Date** 1994 **Origin** Germany

**Engine** 5-cylinder, 120 hp

**Payload** 620 kg (1,367 lb)

Built in limited numbers from 1992–1997, the 290 GD was based on the G-Wagen. It handled well on tricky terrain thanks to 4WD, low gearing, and a locking differential over both axles.

**Load bay** can fold on three sides to serve as a flatbed

Magirus tractor,
*c.*1914, Austria

Key Manufacturers
# The IVECO Story

IVECO is one of the younger truck companies in the market, but it has a long and complex backstory and has produced many award-winning trucks over the years. Created by the merger of five European truck manufacturers, IVECO has since forged its own reputation for quality and innovation.

**THE NAME IVECO** – an acronym of Industrial Vehicles Corporation – first appeared on 1 January, 1975, when the company was incorporated. It was created by the merger of five truck manufacturers: Italian companies FIAT Veicoli Industriali, Officine Meccaniche (OM), and Lancia Veicoli Speciali; Unic of France; and German marque Magirus-Deutz.

While IVECO may be under 50 years old, the companies it was formed from go back much further. For instance, the company that became Magirus-Deutz was founded back in 1866 to make firefighting vehicles, adding trucks to its range in 1910. One of this company's main achievements was the invention of

**Range of IVECO trucks, 1975**
IVECO's initial range, bearing the badges of four of the five manufacturers that made up the company. These were gradually phased out over the following years.

the turntable ladder, which quickly became an indispensable part of fire trucks. Magirus remains a leading producer of firefighting equipment.

The first few years of IVECO's existence were primarily focused on integrating the businesses and rationalizing operations – initially there were 200 basic models and 600 versions, covering a weight range from 2.7–40 tonnes (2.7–39.4 tons), as well as buses and engines.

**IVECO Standard Product Range**
In 1985, IVECO began standardizing all its components, so trucks could be assembled at specialized plants in a huge variety of configurations according to customer needs.

Integration was complete by 1980. At first, trucks were produced under their original brand names, before being phased out and replaced. The first to be substituted for an IVECO-brand truck was the OM Lupetto, which was superseded by the IVECO Zeta range in 1977. A new turbodiesel engine, launched in 1980 and adopted across all ranges, was one of the company's first innovations. In 1984,

the TurboStar, developed with drivers to enhance comfort and performance, became the first heavy commercial vehicle to carry the IVECO name.

In 1986, IVECO formed a joint venture with Ford of Europe's truck division to create IVECO Ford Trucks Ltd, with IVECO holding a 52 per cent stake. The same year saw IVECO acquire ASTRA, an Italy-based manufacturer of heavy-duty trucks. The company's expansion continued in 1990 when it acquired Pegaso, a Spanish manufacturer of industrial vehicles, followed in 1991 by the acquisition of British truck manufacturer Seddon Atkinson.

**Today,** IVECO manufactures **all over the world** and **sells** vehicles in **160** countries.

**Fiat 40**

**Magirus fire truck**

**EuroTech**

**Ecostralis**

**1866** Magirus Kommanditist (Magirus-Deutz) is established
**1899** OM formed
**1903** FIAT produces first commercial vehicle
**1934** FIAT acquires OM
**1937** Lancia Veicoli Industriali starts producing military trucks
**1966** Fiat acquires UNIC (founded in 1905)
**1975** IVECO is formed from merger of FIAT Veicoli Industriali, OM, Lancia Veicoli Speciali, Unic, and Magirus-Deutz
**1977** Launch of first IVECO-branded truck

**1980** Fiat takes ownership of IVECO
**1980** Turbodiesel engine is introduced
**1984** Launch of the TurboStar
**1986** IVECO Ford Truck Europe joint venture is established
**1986** Acquisition of quarry and heavy-duty truck manufacturer ASTRA
**1990** Acquisition of Spanish industrial vehicles manufacturer Pegaso
**1991** Acquisition of UK truck manufacturer Seddon Atkinson
**1991** Launch of Euro range of trucks

**1992** EuroCargo wins International Truck of the Year award
**1992** EuroTech heavy-duty model launches
**1993** EuroTech wins International Truck of the Year award
**1994** IVECO produces first gas-powered truck
**2002** Launch of the Stralis
**2003** Stralis wins International Truck of the Year award
**2011** Fiat spins off non-car activities, including IVECO, into Fiat Industrial
**2012** Stralis Hi-Way launched

**2012** First truck to run on liquefied natural gas launches
**2013** CNH Industrial is founded following the merger of Fiat Industrial and CNH Global NV
**2013** Stralis Hi-Way wins International Truck of the Year award
**2019** Iveco replaces Stralis with S-Way
**2019** Announcement of Nikola Tre FCEV electric truck
**2022** IVECO Group formed after demerger from CNH Industrial

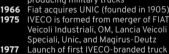

**IVECO-Magirus fire truck features on this 2000 Cambodian stamp**

By this time, IVECO was designing and introducing its own models. In 1991, it announced its new Euro range, covering the entire weight range of trucks. This included the EuroCargo range, which took the International Truck of the Year award in 1992. The EuroCargo is still in production, but has had numerous updates and refreshes over the years.

In 1992, the EuroTech heavy-duty model was introduced, replacing the short-lived TurboTech. This, too, was met enthusiastically by European customers and critics alike. It took the International Truck of the Year award in 1993, marking the first time this coveted award had been won by the same manufacturer over consecutive years. The EuroTech was produced until 2002, when it was replaced by the Stralis. This likewise won the International Truck of the Year title, in 2003. Its successor, the Stralis Hi-Way, launched in 2012, also took the prestigious award in 2013.

In 2019, IVECO began to refresh its range again, with the launch of the heavy-duty S-Way, followed in 2020 by the X-Way, which is aimed at the construction sector.

IVECO has been looking to the future and alternative forms of power since 2012, when it introduced a liquefied natural gas (LNG) version of its Stralis. It now produces LNG and natural

compressed gas versions of all of its vehicles, and is the only major manufacturer to do so.

In 2019, IVECO and heavy-duty EV manufacturer Nikola presented a jointly developed electric truck. Based on the S-Way, it had a range of more than 500 km (300 miles). A hydrogen cell prototype, the Nikola Tre FCEV 6x2, soon followed. Today, IVECO has

**IVECO truck at the Dakar Rally, 2021**
For more than 40 years, IVECO has competed in the gruelling Dakar Rally, winning the truck class on several occasions.

manufacturing plants in Europe, China, India, Russia, Türkiye, Australia, Argentina, Brazil, and South Africa, and has committed to producing net zero carbon by 2040.

# Flying High

Monster trucks are purpose-built, off-road vehicles with supercharged engines that are designed to perform on the most challenging of terrains. Their oversize tyres are typically more than 1.7 m (5.6 ft) high, with a total height of 3.7 m (12 ft). Crowds of thousands come out to watch them race on specially prepared symmetrical tracks with ramps or other obstacles, or to perform stunts, such as the ever-popular vehicle crush, where they squash other vehicles under their huge wheels.

## STEALING THE SHOW

Monster trucks first emerged in the US in the 1970s, when they were sideshows to motocross, tractor-pulling, or mud-racing meets. They quickly gained in popularity and are now often the main event. While monster trucks started out as heavily modified pickup trucks, their design and manufacturing has evolved significantly over the years. They are now mostly custom made with tubular chassis and fibreglass bodies, which are easy to replace if damaged. There are various monster truck series around the world – Monster Jam in the US is the biggest – which include racing and freestyle stunt competitions featuring spectacular jumps and wheelies.

**One of the most famous monster trucks,** Bigfoot has been thrilling audiences around the world since 1975. Here, a current version attempts to clear four cars.

1910 G3 2.5-ton Stake truck from
Reliance, acquired by GM in 1909

# Key Manufacturers
# The GMC Story

While GMC and Chevrolet trucks remain separate brands under the
General Motors banner, their history is deeply intertwined. The saga
of this storied motor company and truck manufacturer is long and
complicated, and began in the early 1900s when brothers Max
and Morris Grabowsky designed their first pickup.

**WHAT IS NOW KNOWN AS GMC**
began with two brothers, Max and
Morris Grabowsky, who founded the
Grabowsky Motor Company in 1900.
They had only built one and a half
trucks when, in 1902, they renamed
the business the Rapid Motor Vehicle
Company after moving to their new
plant on Rapid Street in Pontiac,
Michigan – the future site of GM's
82-acre complex, Pontiac West
Assembly. Rapid trucks were selling
by the hundreds by 1904. Their
1–3-ton models included electric
trucks, in addition to those powered
by a two-cylinder petrol engine.

Meanwhile, William C. Durant had
been building horse-drawn vehicles,
also in Michigan. The millionaire
recognized the progress motorized
vehicles would bring to the transport
industry, so in 1904 he bought a
small company called Buick.

Four years later, Durant and his
carriage-making partner, Robert
McLaughlin, formed the General
Motors Holding Company and
bought the Olds Motor Works, later
known as Oldsmobile. They also
acquired two car makers known for
trucks, Reliance and the Rapid Motor
Vehicle Company founded by the
Grabowsky brothers, which were
combined to form the General Motors
Truck Company in 1911. Reliance and
Rapid trucks were officially rebranded
as GMC in 1912.

**Chevrolet cab assembly line, 1976**
The "Rounded Line" C/K chassis was
manufactured from 1973–87. They were
assembled with newly designed high-tensile
carbon steel ladder-type frames.

**1919 Oldsmobile Canopy Express**

**1937 Chevrolet half-ton pickup**

**1959 GMC 860**

**Chevrolet Avalanche**

| | | | |
|---|---|---|---|
| **1900** The Grabowsky Motor Company (later the Rapid Motor Vehicle Company) is established | **1929** Chevrolet launches International Series AC Light Delivery truck, the first with a closed cab and six-cylinder engine | **1942** Production of trucks halts to produce military vehicles for WWII | **1994** GM's first off-road pickup – the S-10 ZR2 – is designed to have more ground clearance |
| **1908** William Durant co-founds the General Motors Holding Company | **1931** The first factory-built Chevrolet truck is produced | **1947** Launch of Chevrolet Advance-Design and GMC New Design trucks, featuring iconic horizontal 5-bar grille | **2002** The Avalanche debuts as a hybrid pickup truck and SUV |
| **1909** General Motors buys Rapid | **1935** GM opens truck market to everyday drivers with the Suburban Carryall, now the longest-used car nameplate | **1959** GMC's first heavy-duty truck, known as the Crackerbox, is manufactured | **2008** The Suburban HD launches |
| **1911** Durant co-founds Chevrolet | | | **2010** General Motors Ventures is created to advance new technologies |
| **1911** General Motors buys Reliance and forms the General Motors Truck Company, later renamed GMC | **1939** The AC and AF series (the latter with a cab-over-engine design), launch | **1967** Debut of the C/K Series, GM's first to be manufactured on a dedicated truck platform | **2014** Mary Barra becomes GM's first female CEO |
| **1918** The first Chevrolet truck – the Model 490 – is introduced | **1941** The GMC C and E Series light-duty trucks are unveiled | **1971** Chevrolet El Camino is reintroduced | **2022** GMC announces its first fully electric pickup for more than a century |
| | | **1982** Chevrolet S-10/GMC S-15 launches | |

**GMC advertisement, 1980s**
This advertisement shows the full range of 1980s GMC trucks, from its "square-body" third-generation pickups to the heavy-duty Class 8 GMC General.

By that time, Durant had been forced out of the company due to cash flow issues caused by his rapid expansion. In 1911, he convinced race car driver Louis Chevrolet and other financers to invest in a new venture they called the Chevrolet Motor Car Company. Durant and Chevrolet soon parted ways, but the business remained, and the Buick racer's name would go on to become an iconic brand of trucks.

In 1916, Durant used his profits from Chevrolet to purchase enough shares of his former General Motors Holding Company to regain control. When Durant and McLaughlin merged their companies in 1918, the General Motors Company (GM) was born. GM quickly grew, and housing was built for 85,000 workers at its Flint, Michigan, headquarters.

During WWI, 90 per cent of GM's truck production was put towards the war effort. GMC's Model 16 served as the standard US Army ¾-ton truck, of which they made 15,000. They would produce a further 600,000 military trucks during WWII.

Chevrolet introduced its first commerical truck, the Model 490 Light Delivery truck, in 1918. It was based on a car chassis and only rated at a half tonne (0.4 ton), but it was a success and laid the groundwork for the popular pickups to come.

In the 1920s and 30s, GM began to produce trucks for specific markets. The company kept both the Chevrolet and GMC labels for their truck series, with the latter offering upmarket versions of the same models. For instance, the 1947 Chevrolet Advance-Design series of light- and medium-duty trucks had a more expensive counterpart in the GMC New Design series, while the 1955 Chevrolet Task Force sold as the Blue Chip series under the GMC label.

In 1959, GMC brought out its first heavy-duty commercial vehicle, an aluminium cab-over that was so tall and thin, it was nicknamed the "Crackerbox". Their final heavy-duty model, the Class 8 GMC General, was phased out 27 years later, just before their heavy-truck operations were turned over to Volvo in 1987.

In 1960, GMC and Chevrolet launched the first generation of the popular C/K series, which were GM's first pickups built on a truck chassis.

After four generations, these were rebranded to their current nameplates: the GMC Sierra and the Chevrolet Silverado.

GM has weathered many financial downturns in recent years, including recessions in the 1990s and 2008. The latter saw the US government bail out the auto industry. GM made the tough decision to discontinue its Pontiac and Oldsmobile brands, but GMC and Chevrolet survived.

Today, GM is aiming to lead in the growing electric vehicle (EV) market, with the goal of seeing everyone – including truck and semi drivers – in an EV. Their first battery-electric full-size pickup, a green makeover of the Silverado, launched in 2023.

**Reveal of updated Silverado, 2019**
First introduced in 1999, the Silverado replaced Chevrolet's classic C/K range. The light-duty pickup got a major redesign in 2019 to be lighter, yet roomier and more powerful.

" There's a big population that buys our trucks. **It's their life** – or it's their livelihood. *Not their lifestyle, their livelihood.*"

MARY BARRA, GM CEO SINCE 2014

# The Art of the Truck

Around the world, some people spend vast sums to make their trucks stand out from the crowd. They add parts, such as lights, bells, and all manner of chromework, or paint them, with everything from ornate traditional designs to airbrushed cartoon characters. It has become an art form and culture in its own right, and can be seen at festivals around the world.

**Brush strokes**
Trucks such as this are common in the UK and Europe, with ornate and skilful airbrushed artwork added to regular trucks that often work a five-day week.

Trucks have featured artwork since their earliest days when signwriters would paint on company names in ornate letters, along with contact details. What started as a practical measure soon developed into something much more spectacular, as drivers and company owners wanted trucks that were memorable. Today, truck art has become more varied and creative, and it is celebrated at festivals around the world.

In Europe and the US, some operators have designs airbrushed onto the cab. These might commemorate company founders or family members, or feature popular films, TV shows, or sports stars.

Drivers get to show off their trucks at a myriad of events in many countries across the world – from smaller, local gatherings to large-scale festivals. The biggest events include thrilling driving displays and activities in a celebration of trucking culture.

Some attend just to be with like-minded people, but others go to win prizes and will spend hours – if not days – cleaning and polishing their trucks.

Europe's largest annual truck festival is the Truckstar Festival, held at the Assen racing circuit in the Netherlands. Here, more than 2,000 trucks and some 50,000 people from all over the continent converge in late July to celebrate all things truck – and to show off their modified and decorated machines.

## Jingle all the way

In South Asia, drivers in Pakistan, India, and Afghanistan also spend much time and money on making their trucks stand out. The tradition goes back to the 1920s, when Bedford trucks were first imported from the UK. At that time, bumpers and wooden panels along the cab were painted. As long-haul freight developed in the 1940s, businesses designed eye-catching logos, so that people could recognize the truck and who owned it. Soon, this became competitive, and businesses and drivers would make ever-more elaborate designs to out-do their rivals. Truck decoration also harkened back to the Sufi tradition of painting shrines to gain religious favour.

Today, Karachi in Pakistan is a centre for this artwork, with about 50,000 people employed in painting ornate designs in bright colours. As well as artwork, trucks feature ornaments such as bells – hence their nickname of "jingle trucks", given by American servicemen who saw them in Afghanistan.

## All lit up

Another country that takes truck art and customization very seriously is Japan. Here, *dekotora* is a vibrant truck culture that has regular huge gatherings of owners across the country. In short, *dekotora* involves trucks being heavily customized with an array of additional bodywork, including bumpers, bullbars, and wings, as well as impressive lightbars and exhaust stacks – much of which is metal or chrome that is polished for hours before going on display. But it is not just about the lights

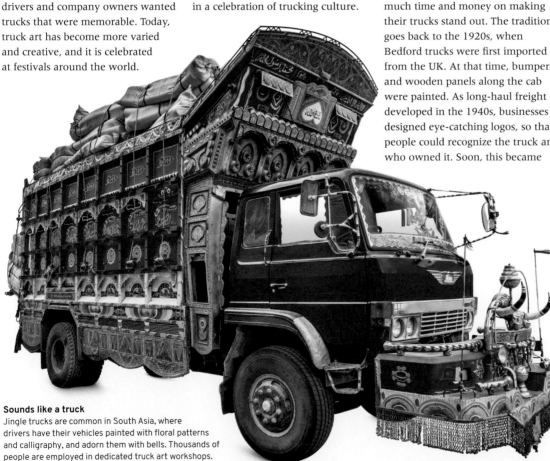

**Sounds like a truck**
Jingle trucks are common in South Asia, where drivers have their vehicles painted with floral patterns and calligraphy, and adorn them with bells. Thousands of people are employed in dedicated truck art workshops.

### Full metal jacket
Resembling something from an Anime show, *dekotora* in Japan takes ordinary trucks and customizes them to extremes. The trucks are so big that gatherings are banned from metropolitan areas.

and chrome – trucks also have ornate artwork painted on their sides and back – often inspired by traditional Japanese art. While some trucks feature contemporary images, such as from anime shows, others are adorned with poetry.

To really appreciate the full majesty of *dekotora*, enthusiasts can attend night gatherings, where the drivers put the lights on. As these bright chrome-bedecked monsters are often banned from urban areas, the regular gatherings take place in more rural settings. Sadly, due to the cost involved in creating dekotora trucks, they are becoming scarce.

Indeed, truck art and customization are not cheap anywhere. In Pakistan, drivers have been known to spend more than a year's salary on decorating their trucks. It is an investment that often pays off, as customers are more likely to remember a truck that stands out from the crowd, and hire it as a result.

## In **Pakistan**, and around the world, **artwork** is seen as an **investment**, as customers are more likely to work with a **beautiful truck.**

PIONEERS

## Haider Ali
One of the most famous truck artists in the world is Haider Ali. Born in Karachi, his father was a master truck painter, and Haider studied under him, painting his first truck at 16. He has developed a distinctive style, and paints anything from landscapes to a truck owner's children. International commissions, include a Bedford truck in Luton, UK, for an exhibition at the Stockwood Discovery Centre in 2011. Luton was where Bedford trucks were made, and they were exported in great numbers to Pakistan in the 20th century.

**Renowned artist Haider Ali** paints a trailer with bespoke designs using oil-based sign paints at his truck workshop in Karachi, Pakistan.

# Toyota Hilux Xtracab SR5

Although the Hilux had been in production since 1968, it was this fourth generation, introduced in 1986, that firmly put Toyota on the international pickup truck map. Known as the Toyota Truck in the US, the Hilux was a success story there. Its fuel injected engine made it more economical to run than rival trucks, and it also proved to be very reliable.

**THE TOYOTA HILUX XTRACAB MODEL** differed from the standard model in offering a small bench behind the front seats, as well as a roll bar with spot lamps. Most of these trucks served long beyond their intended service lives, and gained a reputation for never giving up. While intended as a working vehicle, the Hilux's interior boasted car-like levels of refinement, with comfortable and adjustable seats, air conditioning, and carpeted floors. Compared with engines offered in US-built pickups from the time, the Hilux's R-series' four-cylinder, 2.4 litre engine may seem under-powered. However, it is treasured among Toyota enthusiasts for its durability and dependability.

Pickup body side panel

Fuel filler cap

Off-road tyres

Xtracab side window lets more light into the cabin

1980s equalizer-style graphic

**FRONT VIEW**

**REAR VIEW**

**All in the name**
Founded in 1937, the company name was based on that of its founder, Kiichiro Toyoda. The name Hilux combines the start of "high" and "luxury".

| SPECIFICATIONS | |
|---|---|
| Model | Toyota Hilux Xtracab SR5 |
| Origin | Japan |
| Assembly | Various |
| Production run | 1984–88 |
| Weight | 1,270 kg (2,800 lb) |
| Payload | 955 kg (2,105 lb) |
| Engine | 4-cylinder petrol, 2366 cc, 105 bhp |
| Transmission | 3-speed automatic |
| Maximum speed | Variable |

1
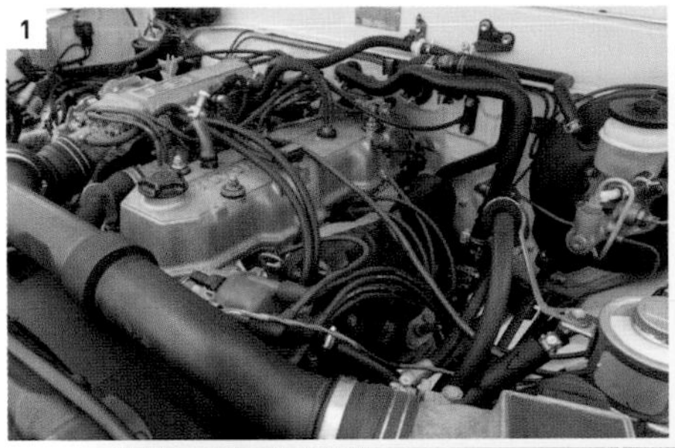

2
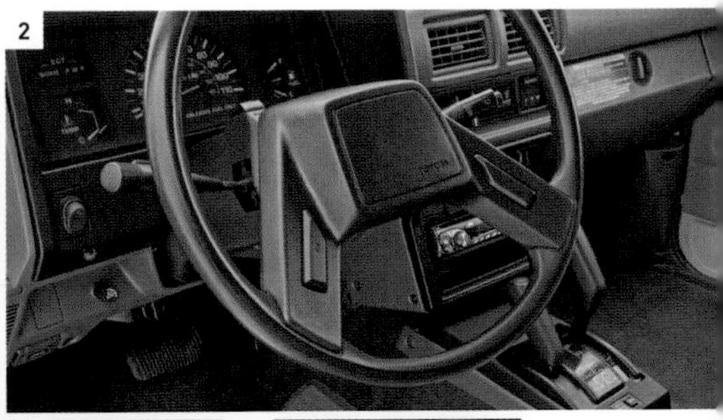

3
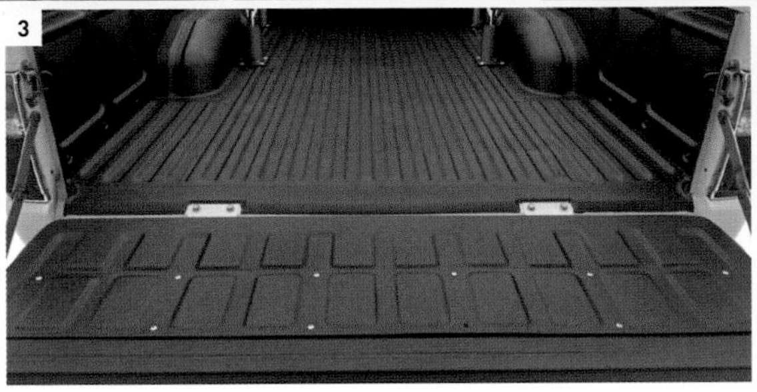

4
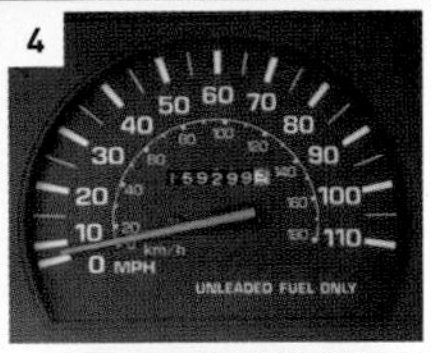

5
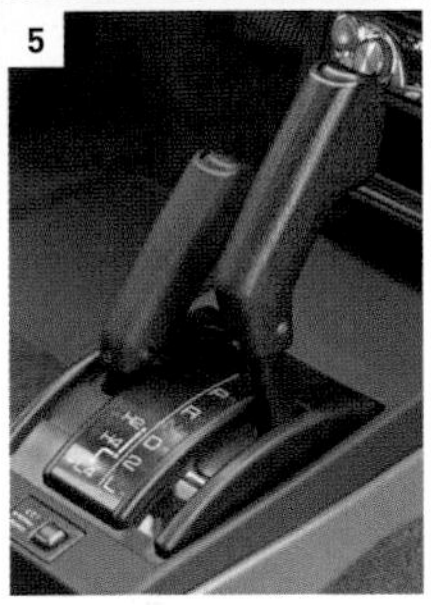

## THE DETAILS

**1.** 2.4-litre fuel injection engine. **2.** Steering wheel and instrument display. **3.** Loading bed with drop-down tailgate. **4.** Speedometer and odometer – this example has very low mileage (500,000 miles (800,000 km) is not uncommon). **5.** 3-speed automatic gearbox with high- and low-range shifter.

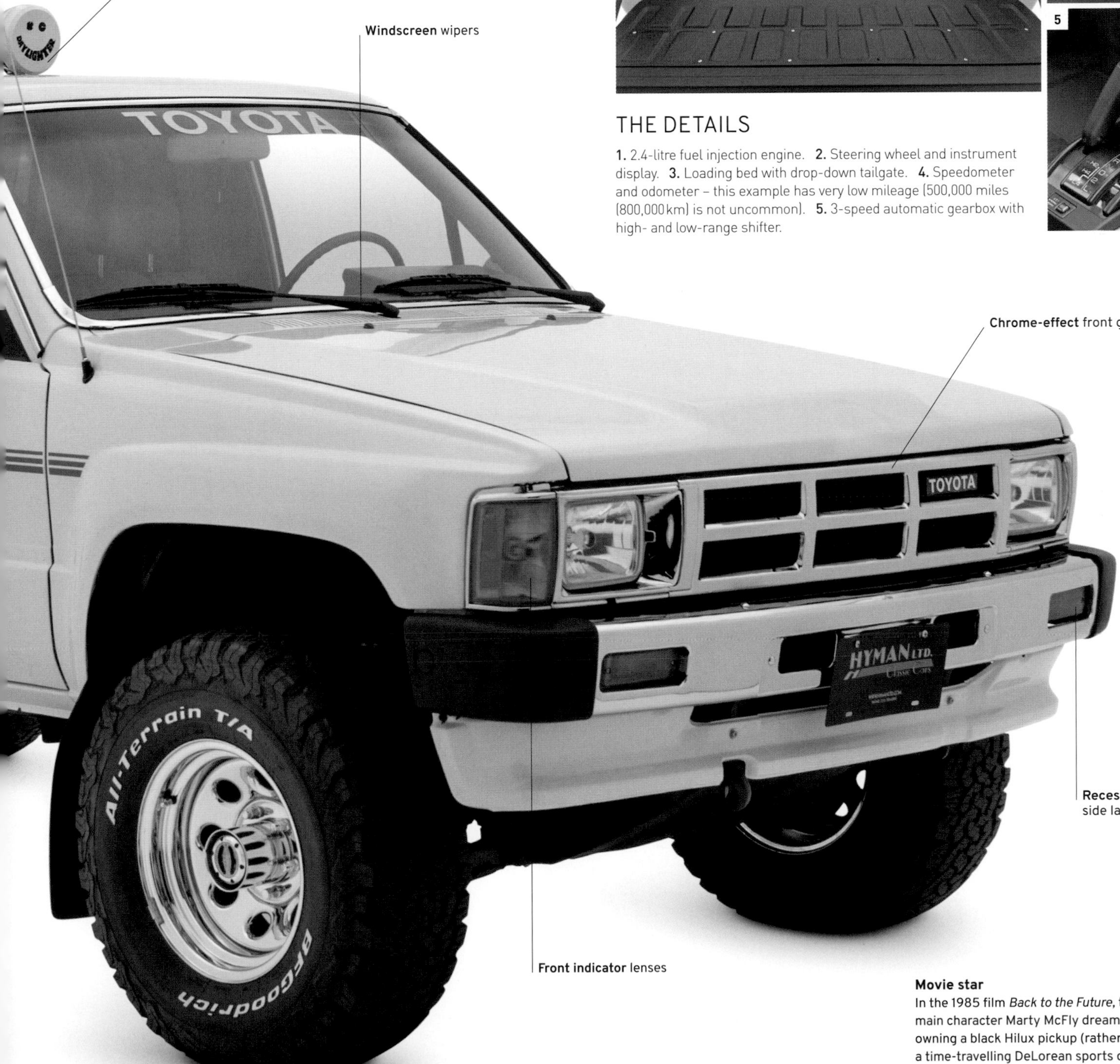
**1980s-style** spot lights
**Windscreen** wipers
**Chrome-effect** front grille
**Recessed front** side lamps
**Front indicator** lenses
TOYOTA
TOYOTA
HYMAN LTD.
All-Terrain T/A
BFGoodrich

Photos courtesy of Hyman Ltd. www.hymanltd.com

**Movie star**
In the 1985 film *Back to the Future*, the main character Marty McFly dreams of owning a black Hilux pickup (rather than a time-travelling DeLorean sports car).

# Aim for the Stars

Moving something as large, heavy, and costly as a fully laden space shuttle required a truly unique vehicle – and that is exactly what Italian heavy-transport manufacturer Cometto provided. Its bespoke, 76-wheel Orbiter Transportation System was built specifically to ferry shuttles at California's Vandenberg Air Force Base. When carrying a shuttle, the vehicle's top speed was 8 km/h (5 mph); unladen, it could reach 21 km/h (13 mph).

## FIRST OF ITS KIND

The 1983 commission from NASA required Cometto to develop and build a giant transporter to carry space shuttles – with payloads installed – on the 27-km (17-mile) trip from the North

Vandenberg Maintenance and Checkout Facility to the launch pad at South Vandenberg. NASA outlined the configuration, but Cometto completed the design details of the 32-m (106-ft-) long transporter – the first of its kind ever developed. Cometto also trained Air Force and contractor personnel in the transporter's operation and maintenance. The transporter was later moved to the Kennedy Space Center, where it remained in use until the end of the shuttle programme in 2011. In 2014, it was bought by SpaceX and used to move Falcon 9 rocket stages.

**The Orbiter Transportation System** carries the *Enterprise* shuttle at Space Launch Complex 6, Vandenberg Air Force Base, California, in 1985.

# Picking Up Speed

This was the age when the might of the muscle car met the practicality of the truck. Before the 1990s, options for trucks that were both fast and practical were limited. But advances in fuel injection and electronic engine management systems allowed manufacturers to produce more cost-effective large-scale performance models that appealed to a younger truck-buying market. Ford offered GT40 supercar engines on a range of its bread-and-butter F-150 trucks, while Dodge dropped its pedigree Viper engine into a Ram pickup. New models were also launched that offered more modest performance improvements. Still enjoyable to drive, they were a more frugal and practical choice for many.

△ **Subaru Brat**

**Date** 1987  **Origin** Japan

**Engine** 4-cylinder, 73 hp

**Payload** 383 kg (844 lb)

The 4x4 Brat was popular with farmers, and were often misused, yet these pickups were reliable. Owners enjoyed their on- and off-road performance. This example has been modified with lowered suspension and modern wheels.

▽ **Lamborghini LM002**

**Date** 1988  **Origin** Italy

**Engine** V12, 444 hp

**Payload** 550 kg (1,200 lb)

Nicknamed the "Rambo Lambo", the 4x4 LM began as a concept named "Cheetah". Many thought the LM002 was a departure for Lamborghini, but the firm had started out building tractors before moving into supercars.

**Raised bonnet bulge** covers fuel-injection system

Photo courtesy of Hyman Ltd. www.hymanltd.com

△ **GMC Syclone**

**Date** 1991  **Origin** USA

**Engine** V6 turbo, 280 hp

**Payload** 226 kg (500 lb)

A wolf in sheep's clothing, the Syclone packed a high-performance engine that took it from 0 to 97 km/h (60 mph) in just 4.3 seconds. It was the fastest pickup on the market and journalists compared its performance to that of a Corvette or Ferrari.

**Angular body lines** give a bold appearance

▽ **Holden Commodore**

**Date** 1995  **Origin** Australia

**Engine** V6, 197 hp

**Payload** 710 kg (1,565 lb)

The Commodore was the first in a new line-up of utility pickups produced by Holden aimed at younger buyers. Options included a new manual gearbox to make use of the power available. The dashboard featured new digital instruments.

▷ **Ford F-150 Lightning**

Date 1995  Origin USA

Engine V8, 240 hp

Payload 337 kg (743 lb)

The Lightning marked a new chapter for the F-150, giving buyers a pickup that performed like a Mustang. The first generation added trim enhancements and bodywork fairings to suggest it was anything but standard.

△ **Dodge Ram 2500 V10**

Date 1995  Origin USA

Engine V10, 304 hp

Payload 750 kg (1,650 lb)

A pickup with an 8-litre Dodge Viper supercar engine. Although it sounded outlandish on paper, the V10 offered true value for those who required extra towing power.

Roof-mounted spotlights

▷ **AM General Hummer H1**

Date 1995  Origin USA

Engine V8 petrol, 200 hp

Payload 900 kg (2,000 lb)

AM General began building the Hummer in 1983 for the US military, but public demand led to the creation of this civilian version in 1992. It was produced until 2006.

CTIS (Central Tyre Inflation System) allows tyres to be inflated/deflated on the move

Custom-made cargo cover

▽ **Ford Falcon Ute**

Date 1999  Origin Australia

Engine V8, 248 hp

Payload 520 kg (1,150 lb)

Holden and Ford dominated the utility vehicle market in Australia for many years. As interest in performance trucks grew, Ford offered this version of its Falcon that went from 0 to 97 km/h (60 mph) in 6.5 seconds.

# Record-breaking Trucks

Trucks are incredibly diverse, and though many are designed for specific purposes, there are some that are truly unique. Manufacturers, designers, tech developers, and even artists have all contributed to creating trucks that have outclassed the rest to claim titles for everything from the most expensive, to the most powerful, to the longest-ever in history.

**BelAZ 75710, 2016**
BelAZ's gargantuan haul truck, being demonstrated at a trade show in Belarus, with a human to show the scale of its eight 4-m (13-ft-) tall tyres. First produced in 2013, the 75710 is the world's largest mining vehicle.

The versatility intrinsic to truck design has resulted in plenty of unusual one-offs. But among production-line models, there is no doubt: the BelAZ 75710 is the biggest truck the world has ever known. It is an off-road haul dump truck used at the largest mines. The 2013 model above, built in Belarus, is in the Ultra Class, which means it must have a minimum payload of 272 tonnes (268 tons). The 75710, however, can handle even more than that: 496 tonnes (488 tons). As the unladen vehicle itself weighs 360 tonnes (354 tons), the fully loaded total is a huge 856 tonnes (843 tons) – roughly 40 per cent more than a fully packed Airbus A380 airliner. Despite this, it can manage 64 km/h (40 mph). Power comes from two 16-cylinder turbodiesel engines each producing 2,300 hp.

The world's largest production semi-truck, by comparison, hits the scales at a road-going 71 tonnes (70 tons), and since 2005 it has been built to order by France's Nicolas Industrie. Its Tractomas TR 10×10 D100 is made to shift extremely heavy oversize loads and measures 126 m (413 ft) in length. One of the few familiar elements is the adapted Renault cab.

## Rich pickings
The most expensive truck ever is a jaw-dropping vehicle called Thor 24, which was built by custom car designer Mike Harrah over seven years at a reputed cost of $10 million. It started out as an ordinary 1984 Peterbilt 359 that Harrah stretched to 13 m (44 ft) in length, in order to accommodate two 12-cylinder Detroit Diesel engines. These are exposed along the truck's long nose, where onlookers can also admire its superchargers. The back boasts a mural of its Norse god namesake. In 2019, this hugely powerful vehicle was sold for $12 million at an auction in Riyadh, cementing its world-record price.

Thor 24 generates 3,424 hp, which should provide exceptional pulling power if it was ever set to work. In the world of catalogued standard production models, however, the most powerful road-going truck hails from China, where in January 2023, Shacman launched its 800-hp X6000 tractor unit. Boasting a 16.6-litre Weichai diesel engine, the truck had 30 hp more than the former record-holder, Scania's 770 hp V8. Before that, Volvo's 750 hp FH16 had ruled for

**Kamaz 4326, 2019**
Russian truck manufacturer Kamaz's 1,000-hp monster races across the Peruvian desert. The 4326 has been a top contender at the famous off-road Dakar Rally since its launch in 1995.

**Thor 24, 2019**
This ultra-customized truck took thousands of hours to build, but it is not all about looks. Thor 24 can reach 209 km/h (130 mph), so long as it then deploys four parachutes to stop.

almost a decade. It seems likely the X6000 will be the last of this powerful kind, given the shift to alternative fuels.

## Endurance race

Power is one thing, but stamina is quite another. No truck brand can claim a record that matches the sheer punishing stamina of Russia's Kamaz. Its 4326 and 43509 military-derived trucks have won their classes on the Paris-Dakar race more than any rival, 15 in total. Despite being top-heavy, its near-50:50 front and back weight distribution, plus the 1,050 hp going to all four wheels, ensures it is almost unstoppable. One driver, Eduard Nikolaev, achieved victories with it in 2013, 2017, 2018, and 2019.

## Going the distance

Unsurprisingly, the record for the longest truck in the world is claimed by the country known for road trains, Australia. In Clifton, Queensland, in

2006, driver John Atkinson piloted a Mack Titan hitched to 112 trailers at a sponsored event. The ensemble measured 1,474.3 m (4,837 ft) in length. While proving this incredible road train could actually move, it was not going anywhere fast, as Atkinson only covered 137 metres (150 yd), painfully slowly.

A more meaningful, if less dramatic, record was set in 2016 by a pioneering autonomous truck built by the tech company Otto. It undertook the world's longest driverless lorry journey by covering 212 km (132 miles) between Colorado Springs and Fort Collins – to deliver 51,744 cans of beer. Nonetheless, that kind of distance would be scorned by William Coe, Jr., of Tallahassee, Florida, USA. In March 2009 he was declared the truck driver with the most distance under his belt in history, covering 4,830,085 km (3,001,276 miles) in 23 years – without a single preventable accident.

## The Big Picture

To celebrate the millennium, Turbozone International Circus – a European circus company known for its pyrotechnic displays – commissioned British artist Banksy to turn a 1988 Volvo FL6 17 tonne (16.7 ton) box van into a mobile painting. The completed makeover was entitled *Turbo Zone Truck (Laugh Now But One Day We'll Be In Charge)*.

It caused a huge stir wherever it went in Europe and South America. Although the basic truck was worth just £1,000, the guide price went up to £1–1.5 million when it was offered at auction by Bonhams in 2019. Even though it failed to sell, it is still likely to be the most valuable piece of "truck art" in the world.

**This truck** features spray-painted images of gears and anvils, along with flying monkeys and a man smashing a TV. The other side pictures soldiers doing a "turbo charge".

# KEEP ON TRUCKIN'

2000–Today

# KEEP ON TRUCKIN'

**The new millennium** brought a huge change in global logistics. Consumer demand for goods increased – not just due to population growth, but also because of the way purchases were made. The growth of online shopping led to a mushrooming of massive, out-of-town distribution centres. This in turn required more and larger trucks to deliver to the centres, and fleets of smaller delivery vehicles to take goods onward to consumers.

Many new trucks now have fully automatic gearboxes as standard, Wi-Fi connection, and LCD screens for mirrors. Fresh methods of powering trucks are rapidly being developed as the need to reduce dependence on fossil fuels becomes ever more pressing. From lithium batteries to hybrid power systems and hydrogen fuel cells, alternative power sources will be the future of trucking.

However, two things have not changed since the pioneering days of trucks at the start of the 20th century. The basic layout of a truck still consists of a cab for the driver and a generous cargo space or a means to pull a trailer. The other element is the all-important human factor: the driver. Trucks still need people who are willing to dedicate a large part of their lives to being on the road serving communities, both locally and globally. There is now an expectation that, at the click of a button, consumers can have goods delivered to their doorsteps, sometimes from another country, the very next day. For the time being at least, a large part of this will be thanks to a truck and a driver.

Up from **57 per cent** in 2000, trucks **now** make up **almost 83 per cent** of **all motor vehicle production** in the United States.

△ **Delivering the goods**
The shift of retail sales from shops to online sites has increased demand for warehousing to store stock. Trucks reverse up to dedicated warehouse bays for quick loading and unloading.

## Key events

▷ **2000** In Australia, Holden launches the Ute (short for utility vehicle), a high performance V6 and V8 pickup.

▷ **2000** Volvo introduces its VHD (Volvo Heavy Duty) truck into North America; it is designed for construction and specialist applications.

▷ **2003** General Motors launches the GMC Canyon/Chevrolet Colorado mid-sized pickup; it is manufactured in Australia by Holden.

▷ **2004** International XT, an extreme sports utility truck built on a medium-sized chassis cab, is produced for the US market.

▷ **2005** Mitsubishi launches its 4th Generation L200 pickup.

▷ **2006** The last Foden truck is built in the UK.

▷ **2015** Ford launches an aluminium bodied F-150. The F-150 has been Ford's most successful vehicle since its Model T.

▷ **2022** Ford F-150 Lightning electric pickup launches.

△ **Ford F-150 Lightning electric pickup**
Part of the 14th generation of the Ford F-series, the Lightning – shown here at the International Auto Show in New York City (2022) – has a range of 370 km (230 miles).

▷ **2022** Volvo Trucks showcases a hydrogen fuel cell that could power a zero-emission electric truck in the future.

◁ **Futuristic look** German industrial designer Luigi Colani produced this adaptation of a DAF truck in 2000. By giving it an aerodynamic shape and fine-tuning the engine he claimed to have reduced its fuel consumption by 30 per cent.

# International LoneStar

International has been building trucks since 1907, with their current range topped by the LoneStar, a heavy-duty tractor unit. Exported around the world, the LoneStar traverses ice roads in the Arctic Circle, hauls road trains through Australia's outback, and even – as shown here – recovers buses in the UK.

**THE LONESTAR IS PERHAPS** one the boldest designs in modern trucks, blending the past and present with an impressive grille and bonnet. The chassis is available in a range of options, but central to them all is the 6x4 semi-tractor. Recovery trucks, or wreckers, like this one are custom designed and built to the requirements of the operator. A conventional semi-tractor truck chassis is typically lengthened and a recovery wrecking body is then built around it. Other LoneStar trucks that have been adapted in this way include lavish motorhomes and race-car transporters.

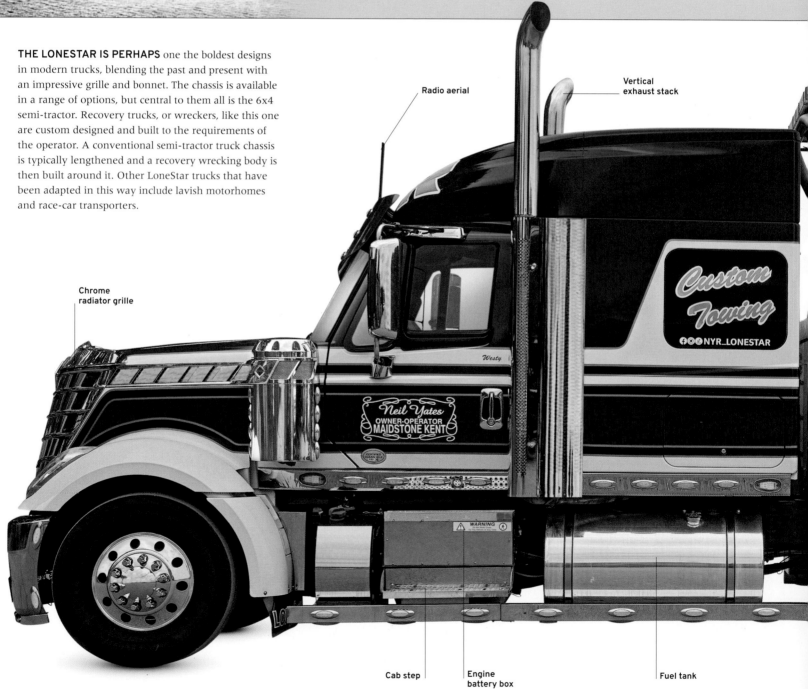

Radio aerial

Vertical exhaust stack

Chrome radiator grille

Cab step

Engine battery box

Fuel tank

**FRONT VIEW**

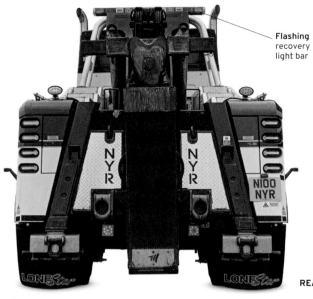

Flashing
recovery
light bar

**REAR VIEW**

| SPECIFICATIONS | |
|---|---|
| Model | International LoneStar |
| Origin | USA |
| Assembly | Springfield, Ohio, USA |
| Production run | 2009–present |
| Weight | variable |
| Payload | variable |
| Engine | 16-litre, 500 hp |
| Transmission | 18-speed |
| Maximum speed | unknown |

**International LoneStar**

To recover a broken-down vehicle, the truck uses the hydraulic lifting arm at the rear to raise the stranded vehicle's front steering axle off the ground. It can then be towed away under the control of the wrecker. If the recovered truck was towing a trailer, the trailer's brake and electrical couplings can be connected directly to the wrecker.

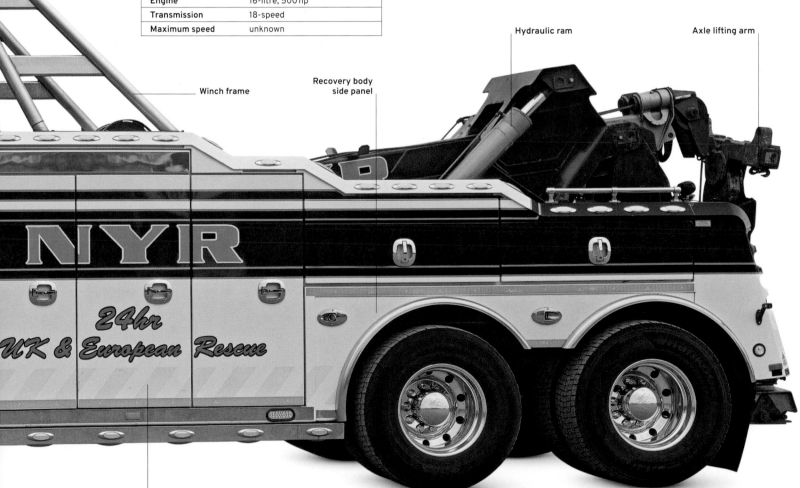

Winch frame

Recovery body
side panel

Hydraulic ram

Axle lifting arm

**Storage locker** for
recovery equipment

## THE EXTERIOR

One of the most aerodynamic trucks in its class, this wrecker version of the LoneStar operates as part of a fleet of heavy recovery trucks offering 24-hour assistance. Flashing lights and amber running lights, highlighted with chrome, not only make this vehicle highly visible to stranded drivers and passing traffic, but also add a dash of style. The rear bodywork is custom-made, with built-in storage lockers that contain vehicle lifting strops and the controls for the hydraulic lifting arm, some of which can be remotely controlled. The large radiator grille, with its Art Deco feel, is a distinctive feature of International trucks.

1. International grille badge  2. Chrome bumper LoneStar logo  3. Cab door handle  4. Flush-fitting front lights
5. Side light  6. Rear LED lights  7. Footplate for access to winch  8. Winch drum and cable  9. Chassis plate giving vehicle weights  10. Tool, ratchet strap, and lifting strop locker  11. Front wheel with chrome wheel cover
12. Hydraulic ram to lower recovery lifting arm
13. Control levers for operating hydraulic systems

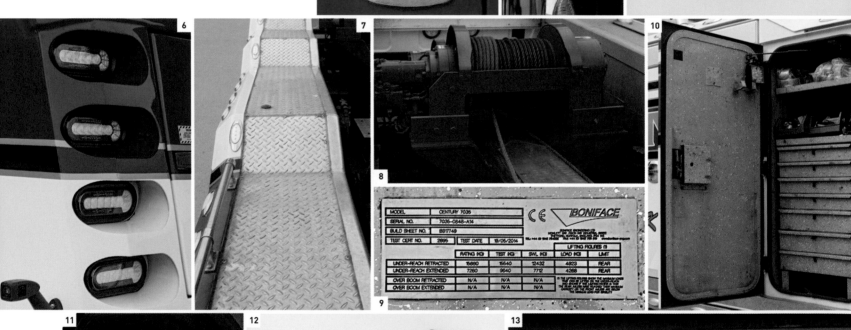

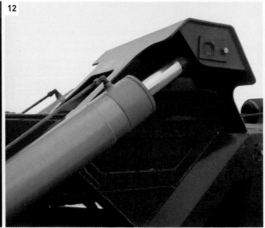

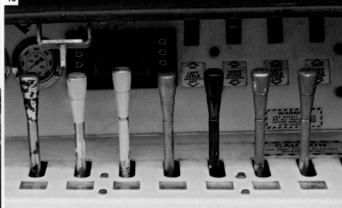

## THE INTERIOR

Inside, this high-rise sleeper cab is trimmed with red leather to match the livery paintwork of the exterior. All the cab controls are angled towards the driver, reflecting the importance of ergonomics in cab design. The transmission is fully automatic and controlled from the switch selector on the centre console. Being a sleeper cab, the driver's area features a full-size double bed – and an impressive custom-built sound system.

**14.** Luxurious cab interior   **15.** Steering wheel and instrumentation   **16.** Cab controls, radio, and the red-and-yellow switches operate the pneumatic parking brakes for the truck and trailer   **17.** Double bed and sound system speakers behind front seats

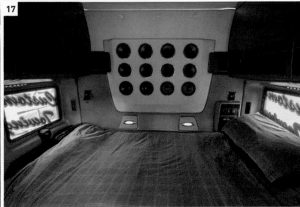

## THE ENGINE

Used in many other tractor units, including those of Kenworth, Peterbilt, and Volvo, the engine is a Cummings ISX. Available in power options ranging from 420–620 hp, it is managed by a host of electronic systems for greater efficiency. Transmission is via an automatic 18-speed gearbox. With the large bonnet lifted, the engine bay is easily accessible.

**18.** Cummings ISX 16-litre diesel engine   **19.** Turbocharger (bottom left) and radiator coolant tank (top right)

# Monster Dump Trucks

As demand for products has increased, so has the need to dig for more raw materials, from stone and coal, to minerals including iron ore, lithium, and gold. The massive machines required to transport a constantly moving landscape of resources grew out of open-mining operations. Modern dump trucks represent the cutting edge of extreme truck development, ranging from those used in construction projects, to others that have become so vast that they require computer-aided driver controls in order to operate them safely.

### ▷ Caterpillar D350

**Date** 2000 **Origin** USA

**Engine** 6-cylinder diesel, 340 hp

**Payload** 28 tonnes (28 tons)

At the smaller end of the scale, the 9.9 m (32.5 ft) articulated D350 is widely used around the world on contruction projects. It can be transported by road on low-loaders.

**70-degree dump angle** allows rapid emptying

### ◁ Volvo A40D

**Date** 2004 **Origin** Sweden

**Engine** 6-cylinder diesel, 420 hp

**Payload** 38 tonnes (37 tons)

Volvo has been making dump trucks since the 1950s. The 6x6 A40D featured many Volvo truck components. It is articulated, which gives it greater manuverability and the hydraulically powered tipper can be raised in just 12 seconds.

## The Blue Miracle

This huge, now-derelict mining machine was built in East Germany in 1964 for excavating coal. At 3,850 tonnes (3,910 tons), 171.5 m (370 ft) long, and 50 m (190 ft) high, it claims to be the largest abandoned machine in the world. Known as "Bagger 258", it moved on steel tracks, while its rotating buckets could plough 15 m (50 ft) into the ground. It had over 50 years of continued use before being retired in 2003.

**The Blue Miracle's** rotating buckets could each scoop up 1.5 sq m (53 sq ft).

### △ Liebherr T 284

**Date** 2012 **Origin** USA

**Engine** V20 diesel, 3,650 hp

**Payload** 363 tonnes (357 tons)

Over 15 m (48 ft) long, this was one of the largest dump trucks. It was built for heavy-scale mining (copper, iron ore, gold) and used across the world. Like many machines of this size, it was shipped in component form and assembled on site.

**Tipper can carry** up to 840.5 tonnes (830 tons)

**75710**

**Height** of 8.6 m (28 ft 3 in)

△ **BelAZ 75710**

| | | |
|---|---|---|
| **Date** 2013 | **Origin** Belarus | |
| **Engine** Two V16 diesels, each with 2,300 hp | | |
| **Payload** 450 tonnes (440 tons) | | |

Over 20 m (67 ft) long, the world's biggest dump truck costs from $6 million and is designed to operate in temperatures of -60 °C to +50 °C (-76 °F to +122 °F). Drivers are trained to operate them on simulators, much like airline pilots.

▽ **Komatsu HD465-7E0**

| | |
|---|---|
| **Date** 2014 | **Origin** Japan |
| **Engine** Diesel, 715 hp | |
| **Payload** 55 tonnes (54 tons) | |

One of Komatsu's smaller mining dump trucks, it weighs 43.1 tonnes (42.4 tons) empty. Its seven-speed electronic transmission with conventional selector makes this truck very easy to operate.

**Front wheels have** 8.5 m (27.9 ft) turning radius

**Overhang** protects the cab from rock and soil

△ **Caterpillar 797F**

| | |
|---|---|
| **Date** 2017 | **Origin** USA |
| **Engine** 20-cylinder diesel, 4,000 hp | |
| **Payload** 363 tonnes (357 tons) | |

Another of the world's biggest dump trucks, this is also assembled on site, requiring up to 13 semi-articulated trucks to transport all of the parts. Double rear wheels provide extra traction, with each one costing over $50,000.

**Ice road trucking**
Driving over frozen roads in harsh winter conditions requires extreme skill and caution. It also demands a suitably rugged truck, such as this Peterbilt.

# Going to Extremes

Trucks are required to travel all corners of the world, including inhospitable places, such as the Sahara Desert and the Alaskan tundra. These trucks and drivers take on the most challenging and dangerous conditions to ensure that their cargo makes it to and from some of the most remote locations on Earth – and gets there on time.

The James Dalton Highway, which stretches 666 km (414 miles) across the Alaskan tundra, is renowned for being one of the most dangerous roads in the world. But as the only route and supply line to the Arctic Ocean and the oil fields there, hundreds of trucks traverse it every day. With stretches having perilous names such as "Avalanche Alley", the Highway can be dangerous for even

experienced truck drivers, and is regularly affected by snowfall and ice during winter, and flooding in the warmer months.

The TV show *Ice Road Truckers*, (2007–17), highlighted the dangers drivers face on these treacherous roads. Driving through snow, and where temperatures can reach -80°C (-112°F), their trucks have huge fuel tanks as there are no fuel stops on the

route. As the name suggests, they sometimes drive on frozen lakes and rivers. The US Department of Transport test and approve routes to run on, which is cheaper and easier than building permanent roads and bridges. However, these routes often have speed limits of about 32 km/h (20 mph), which

**Don't look down**
Drivers in the Chinese Himalayas need bravery as well as skill to safely traverse the narrow and dangerous mountain roads.

## Extreme Motorsports

In addition to working in extreme conditions, many trucks are tested during extreme sports, such as the Dakar Rally. The annual event, first held in 1979, was originally run from Paris in France to Dakar in Senegal, but now takes place in Saudi Arabia, having also been held in South America for a spell. Trucks – either series production or modified – race on the same course as the cars and bikes that also compete, covering thousands of miles of rough tracks over the 16 days of the event.

**Paris–Dakar Rally, 1986**
A Spanish Pegaso truck makes its way through Algeria in the 1986 race.

can make journeys slow. The trucks are kitted out with everything drivers need, including ways to keep warm if they suffer a breakdown en route.

### Storming the desert

At the other extreme, trucks hurtle through the deserts of Africa, getting goods to remote areas. These trucks go over roads that are sometimes barely more than sand, operating in intense heat.

The chances to stop and get repairs are few and far between. Despite that, people will often hitch a lift with a truck as it can be one of the only ways to travel long distances.

Conditions are similarly tough in the Australian Outback, where temperatures that can hit more than 50°C (122°F) and there is often hundreds of miles between towns. But again, there is a dedicated band of truck drivers who pilot tractor units pulling 175-tonne (172-ton) road trains that are more than 50 m (164 ft) in length across this barren land – and love it.

Powered by 700-hp-plus engines, the tractors are fitted with bull bars at the front, in case an errant kangaroo or cow gets in the way. They are also fitted with ice packs to protect the driver in case of a breakdown or an unexpected stop, and special cooling features to manage high-risk items, such as batteries and differentials in the extreme heat.

With such lonely conditions, drivers have to fend for themselves – if they experience a tyre blow-out, there are no breakdown trucks nearby to help, so they must change it themselves.

### Treacherous paths

Meanwhile for some drivers, such as those traversing the Chinese Himalayas on roads like the Sichuan-Tibet Highway, a head for heights is a prerequisite. The highway, which stretches more than 2,000 km (1,200 miles), crosses 14 mountains that average 4–5,000 metres (13,100–16,400 ft) in height. It is often narrow, with tight, twisty corners.

As a further challenge, weather conditions can change quickly and landslides and rockfalls are common. Drivers know that one mistake could spell disaster, and wrecked vehicles are a common sight along the route.

### Pioneers

Trucks and their drivers have been overcoming extreme conditions for decades, providing a vital supply route to remote locations. In 1964, a couple of intrepid British truck operators opened the first commercial truck route between Western Europe and the Middle East. Friends Bob Paul and Michael Woodman left the UK in spring 1964 and arrived in Kabul, Afghanistan, 31 days later.

They made the journey with only basic food supplies, and small-scale relief maps and a compass. While it may have been dangerous – many roads were not properly developed – the trip was a success and it paved the way for thousands of other trucks to make similar journeys.

So why do drivers put themselves through such gruelling conditions? Many have no choice, as these treacherous routes are the only ones available to them. For others, it is money, as driving in extreme and dangerous conditons commands a premium pay. Some simply enjoy the thrill and camaraderie. That shared experience of driving in the toughest, most dangerous of conditions binds them together. Often, if one truck has stopped, another driver will stop and check on their wellbeing.

**Pile them high**
Trucks can be adapted with high wheel rims to aid desert driving. They are often overloaded because deliveries to inhospitable terrain, such as the Sahara below, can be infrequent.

## Some trucks have fuel tanks with **more than 1,800 litres capacity** as they can **travel more than 800 kilometres** before passing another fuel stop.

# GMC Hummer EV

The first GMC Hummer – or "Humvee" – was a legendary vehicle designed for the US military in 1983. At 2.2 m (7 ft) wide, it had the dimensions of a truck and a versatile chassis. It also lasted years beyond its expected service life. In 1992, a civilian station wagon version was launched. Today's fully electric EV model introduces the latest technology to give a range of up to 529 km (329 miles) and a power output of up to 1,000 hp.

**ONE OF THE HUMMER EV PICKUP'S** most radical features is its "CrabWalk" four-wheel steering, which allows it to drive diagonally. It also has a launch control feature that helps it get from 0–96 km/h (60 mph) in 3.3 seconds – acceleration not generally seen on a truck. Packed with new technology, it has a wealth of cameras and sensors to inform the driver of potential hazards, and in "extract mode" (which raises the air suspension up to maximum height) it can ford through water up to 81 cm (32 in) deep. With three electric motors producing a combined 100 hp, it is capable of towing 3.4 tonnes (3.7 tons).

1.5 m (5 ft) loading area

89 cm (35 in) mud terrain tyres

Side steps make entry and exit easier

**FRONT VIEW**

**REAR VIEW**

**Hummer logo**
General Motors have designed a new logo for the electric Hummer to differentiate it from the conventional model.

## SPECIFICATIONS

| | |
|---|---|
| Model | GMC Hummer EV |
| Origin | USA |
| Assembly | Unknown |
| Production run | 2022–present |
| Weight | 4.1 tonnes (4 tons) |
| Payload | 590 kg (1,300 lb) |
| Engine | 746 kW electric |
| Transmission | Single speed |
| Maximum speed | 171 km/h (106 mph) |

**Roof bars** for carrying equipment such as bikes, kayaks etc

**Bonnet-mounted** ditch lights

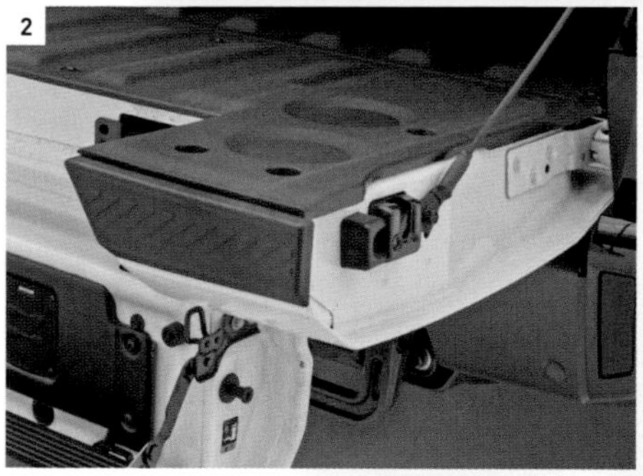

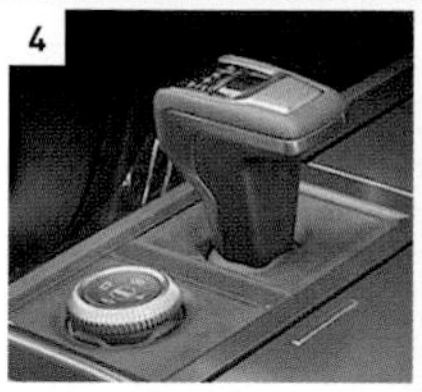

**Lights** extend across the front panel, highlighting Hummer logo

## THE DETAILS

**1.** Built-in sound system control and extra speakers in drop-down tailgate **2.** The tailgate incorporates a down step **3.** Steering wheel and digital instrument panel **4.** Electronic gear selector lever and driving-mode controls **5.** Main vehicle control panel screen

**Front bumper**

**Military-inspired** tie-down eyes used for recovery

**Cooling inlet area** for electrical systems

## Multiple motors

The EV has three electric motors – two at the rear under the floor as one unit and another at the front. Power comes from a 212 kWh battery. Under the bonnet, the "engine bay" provides storage. True to Hummer tradition, the EV retains a large centre console inside and generous rear cabin space.

**Hummer logo** on wheels

Photos courtesy of Hyman Ltd, www.hymanltd.com

# Pick of the Crop

Pickup truck racing has become an established part of the US motor racing scene over the past 40 years. Today, it is part of the NASCAR set-up, boasting grids of 36 pickups and some big-name drivers. Pickup truck racing goes back to 1983, when former NASCAR driver Buck Baker established the National Pickup Truck Racing Association to help graduates of his driving school to start their careers.

### ESTABLISHED SERIES

It was not until 1995, however, that a pickup truck series for NASCAR debuted (although exhibition races were run during the previous year). It is now one of three national series sanctioned by NASCAR, alongside the Cup Series and Xfinity Series. Showcasing pickups that are popular in the US, races most often take place on oval circuits, including iconic tracks such as Daytona and Talladega, with a set number of laps. Other countries also have their own pickup truck racing series, including the UK, Mexico, Brazil, Australia, and Thailand, all of which run along similar lines to the US.

**NASCAR Camping World Truck Series** A Chevrolet Silverado 250 race held at Alabama's Talladega Superspeedway in 2022.

# Modern Rigs and Tractor Units

The biggest recent advance in modern trucks has been the inclusion of sophisticated electronics and automatic gearboxes as standard features. Drivers are now aided by digital-screen mirrors and reversing cameras, while sirens warn cyclists of the truck's movements. Trucks are becoming increasingly car-like to drive – a far cry from the pioneering days of trucking at the start of the 20th century. Facing rising fuel and materials costs, and environmental concerns, manufacturers are developing alternative power sources, including electric, biodiesel, and hydrogen fuel cells.

▽ **Kenworth W900**

| | | | |
|---|---|---|---|
| **Date** 2000 | **Origin** USA | | |

**Engine** 6-cylinder, 500 hp

**Payload** 36.3 tonnes (35.7 tons) GVW

The large sleeper cab of this 13-speed Kenworth features a double bed, sink, and table. In production since 1961, with many significant upgrades, the W900 is seen by many as the star of the Kenworth range.

△ **International LoneStar**

**Date** 2014 **Origin** USA

**Engine** 16 litre, 500 hp

**Payload** Unknown

The LoneStar is International's largest on-highway truck and its flagship model. The example above has been adapted as a wrecker. Like semi-truck LoneStar models, it has a sleeper cab.

▽ **International ProStar**

**Date** 2016 **Origin** USA

**Engine** 6-cylinder, 450 hp

**Payload** Variable

International trucks are among the most familiar sights on US roads. The ProStar is designed primarily for long-distance highway use, transporting containers, tankers, and low-loader trailers. It is offered with either a sleeper cab or a day cab (shown here).

**Side wind breaker** pushes airflow away from sides of trailer

**Side panel** shaped to improve aerodynamics

**Rear light panel**

△ **DAF XF 480 FTP**

**Date** 2017 **Origin** Netherlands

**Engine** 6-cylinder, 480 hp

**Payload** 44 tonnes (43 tons) GVW

The XF, the largest DAF model, has its second helper axle partly hidden behind the side fairing for a cleaner look. The wheels on this axle are smaller. They are only in contact with road when the load requires it, reducing fuel consumption.

**Roof-mounted** breaker reduces wind resistance over trailer

**Batteries** located both sides of the chassis

ZERO TAILPIPE E

## ▽ Scania 650

**Date** 2019 **Origin** Sweden

**Engine** V8 diesel, 650 hp

**Payload** 44 tonnes (43 tons) GVW

A V8 diesel engine producing 3,300 Nm of torque, and a spacious cab featuring plush trim with lots of storage space, make the 650 well-favoured by drivers and haulage contractors alike.

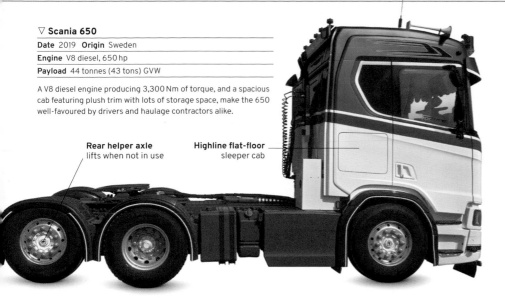

**Rear helper axle** lifts when not in use

**Highline flat-floor** sleeper cab

## ▷ Renault T High

**Date** 2022 **Origin** France

**Engine** 6-cylinder, 480 hp

**Payload** 44 tonnes (43 tons) GVW

T High trucks have been developed for long range logistics. Like the tall Renault Magnum, the T High has a flat floor, giving the driver step-free access to the rear bed. This truck has been custom-fitted with a row of roof-mounted spotlights.

## ▷ Mercedes Actros

**Date** 2022 **Origin** Germany

**Engine** 6-cylinder, 480 hp

**Payload** 44 tonnes (43 tons) GVW

Launched in 1997, the Actros range now sports a new GigaSpace cab, which provides more interior space. The Actros has proved popular with haulage companies and retailers that operate large truck fleets.

**Roof-mounted** air horns

**Adjustable wind breaker** on roof

**Fold-out** side wind deflectors

100% ELECTRIC

## ◁ Volvo FM Electric

**Date** 2022 **Origin** Sweden

**Engine** Electric motors, up to 666 hp

**Payload** 44 tonnes (43 tons) GVW

Zero emissions make the FM ideal for use in urban settings. Six battery packs powering three electric motors give it a range of up to 300 km (186 miles), so longer haul is an option too. The charge time can be as little as 2.5 hours.

**Bold** promotional graphics

## △ Volvo FH13 Globetrotter

**Date** 2023 **Origin** Sweden

**Engine** 6-cylinder, 500 hp

**Payload** 44 tonnes (43 tons) GVW

Volvo's top-of-the-range heavy-haulage truck, the FH can have up to a 750 hp engine, one of the most powerful available. It is widely used in small fleets and by owner/driver operators.

# Promotion in Motion

The Tour de France is one of cycling's most iconic events and runs for three weeks each July in challenging stages across the country. Every year, millions of spectators watch hundreds of riders compete, although it is not the only attraction. *La caravane publicitaire* is a parade of some 150 vehicles advertising a diverse range of brands. It precedes the race and has become a must-see event in its own right.

## CAVALCADE OF COLOUR

*La caravane publicitaire* began in the 1930s as a way to raise more revenue for the Tour by having advertising on signs attached to standard cars. This evolved into companies commissioning custom-made trucks that look more like carnival floats, in a variety of weird and wonderful designs, as each tries to outdo its rivals. Being included in the parade is a major honour that can help boost a brand's profile in France and beyond. Today, the free-to-attend event has become a huge spectacle. The vehicles form a convoy that is about 12 km (7.5 miles) long and takes 45 minutes to pass through, with many participants distributing products, snacks, and novelty souvenirs to people lining the route.

**These chicken-themed trucks** were built to advertise *Le Gaulois*, a well-known French poultry brand, in the 2017 *caravane publicitaire*.

Vabis's first truck, designed in 1902

# Key Manufacturers
# The Scania Story

Scania is an iconic name in European truck manufacturing, and one of its longest established marques. The company is respected for its V8 engines, as well as for pioneering a modular approach to truck production. It continues to produce some of the most powerful trucks available in Europe.

**SCANIA TRACES ITS HISTORY** back to the founding of Vabis in 1891 as a partnership between engineer and industrialist Philip Wersén, and Surahammars bruks, a long-established ironworks.

The company manufactured wagons, horse-drawn trams, and carriages before beginning to develop motor vehicles in 1898.

**The Scania-Vabis badge, used from 1911–54**

In 1900, a newly formed company, Maskinfabriksaktiebolaget Scania, commonly known as Scania, began operations by acquiring the Swedish subsidiary of British bicycle maker Humber & Co. Soon after, it diversified into making gears and cars and in 1902, its first truck, which had a 12 hp engine and could carry 1.5 tonnes (1.5 tons). Vabis also built its first truck in 1902, with a capacity of 1.5 tonnes (1.5 tons). But by 1911, Vabis was failing and merged with Scania to create Scania-Vabis, establishing its head office in Södertälje in the Swedish province of Scania, where it is still located. While Scania-Vabis initially built luxury cars, after World War I it decided to focus exclusively on trucks.

The company struggled at first, due to the influx of cheap demilitarized trucks onto the market. In 1921, it received extra funding to avoid bankruptcy, after which Scania-Vabis began to thrive – and innovate.

In 1939, Scania-Vabis introduced the "Royal" unitary diesel engine, which marked a milestone as the start of Scania's "modular philosophy" of production. The engine was designed so that it could be manufactured from standardized components. This meant that parts could be easily interchanged between models, keeping costs down.

Continuing to innovate with its engines, in 1949 Scania introduced direct fuel injection to its diesel engines, increasing power while reducing fuel consumption. In 1961 it launched the first supercharged truck engine, the 10-litre DS10, which was fitted in 75 series trucks.

**Airflow essentials**
Wind-tunnel testing is a key part of Scania's truck design, ensuring cabs are aerodynamic. In 2013, the company opened a wind-tunnel facility that also emulates climatic conditions.

In 1969 came the introduction of Scania's revered V8 engine. At its launch, the 14-litre turbocharged engine was the most powerful in Europe, delivering 350 hp. It also had a high torque output at low speeds and low revs, which improved fuel economy and prolonged the life of the engine. In the same year, Scania-Vabis merged with Swedish aircraft and car manufacturer, Saab, to form Saab-Scania, with their trucks bearing the Scania brand.

In 1980, Scania applied its revolutionary modular concept to their truck manufacturing as a whole on its GPRT range. Here, engines, gearboxes, and drive shafts were made on a modular basis, but, significantly, so were components

**Trucks for all purposes**
Scania-Vabis' L36 was launched in 1964 to cater to the urban distribution market. It had a turbocharged engine and a payload of 5–6 tonnes (4.9–5.9 tons).

**Type 1444**

|  |  |
|------|-------|
| **1891** | Founding of Vabis, which initially makes wagons and carriages |
| **1900** | Maskinfabriksaktiebolaget Scania is founded – a bicycle maker |
| **1902** | Vabis and Scania both manufacture first truck in the same year |
| **1911** | Vabis and Scania merge to form a single company – Scania-Vabis |
| **1939** | Debut of first unitary diesel engine, known as the "Royal" |
| **1949** | Introduction of direct fuel injection into diesel engine |

**N12**

|  |  |
|------|-------|
| **1957** | Brazilian subsidiary of Scania-Vabis, Brasil S.A., is established |
| **1961** | Launch of supercharged DS10 engine |
| **1969** | Introduction of legendary V8 engine |
| **1969** | Scania-Vabis merges with Saab to create Saab-Scania; Scania brand name appears on trucks |
| **1976** | Production facility in Argentina opened |
| **1980** | Modular concept of truck production is introduced with GPRT range |
| **1991** | Aerodynamic Streamline cab range is introduced |

**T Cab**

|  |  |
|------|-------|
| **1993** | Scania's retarder introduced to automatically control downhill speed |
| **1995** | 4-series launches, winning International Truck of the Year title in 1996 |
| **1995** | Scania and Saab split into separate companies |
| **2000** | One millionth Scania truck rolls off production line |
| **2004** | R-series launches |
| **2005** | Production of T-series bonneted trucks ends |

**R410**

|  |  |
|------|-------|
| **2008** | Volkswagen acquires majority stake in Scania |
| **2009** | Refreshed R-series takes International Truck of the Year title |
| **2014** | Scania becomes wholly owned subsidiary of Volkswagen |
| **2016** | New Generation launches |
| **2019** | AXL, a cabless autonomous concept truck, debuts |
| **2020** | First battery-electric truck launches |
| **2020** | Launch of 770S – the most powerful truck in Europe |

including axles, frames, and cabs. This meant Scania could tailor its vehicles to specific demands while making production more economical.

Further advances came in 1991 with the Streamline cab range, which reduced air drag and improved fuel economy. This proved popular with operators and drivers alike. Scania's attention then turned to driver aids. In 1993, the Scania Retarder was introduced, which automatically controlled downhill speed, reducing the use of wheel brakes by up to 75 per cent.

In 1995, Scania's 4-series was launched with the help of Italian industrial design firm Bertone. With their clean, curved lines, these trucks have come to be regarded as a classic in terms of styling. The PRT range came out in 2004, with different styles of cab identified by each letter: P for small cab-over, R for large cab-over, and T for conventional. In 2005, Scania ceased production of its conventional trucks, with the cab positioned behind the engine, due to declining sales. However, since then, these trucks have developed a cult appeal in the truck community.

**Chassis assembly at Södertälje, 2019**
A Scania S-series truck is built at its factory in Södertälje. The model's modular design allows for it to be easily customized with different engines, gearboxes, and axle configurations.

Scania launched a major new range of trucks in 2016, including the existing R-series and new S-series. The latter came with a flat floor, but both cab designs focused on driver comfort. Three years later, Scania showcased its latest autonomous vehicle concept, the AXL, a heavy self-driving truck without a cab.

The following year, it launched the latest iteration of its V8 diesel engine and its first battery-electric trucks. However, these are not Scania's only alternative fuel trucks. The company has been developing alternative fuel options for more than 40 years and is the only manufacturer to offer every type of energy option: Diesel, HVO (Hydrotreated Vegetable Oil), FAME (Biodiesel), Compressed Natural Gas, Liquified Natural Gas, PHEV (Plug-in Hybrid), and BEV (Battery Electric), and is also trialling Hydrogen Fuel Cell technology.

Scania produced its **one millionth vehicle** in 2000, just 13 years after its **500,000th** had rolled off the production line.

# Volvo FH13 Globetrotter

Volvo won Truck of the Year in 1994 with its FH Globetrotter. The truck was designed for long-distance haulage, where larger cabs are welcomed by drivers who have to spend extended periods on the road. The Globetrotter is now Volvo's flagship truck, rivalling Scania and Mercedes in European cab-over markets.

**FRONT VIEW**

**REAR VIEW**

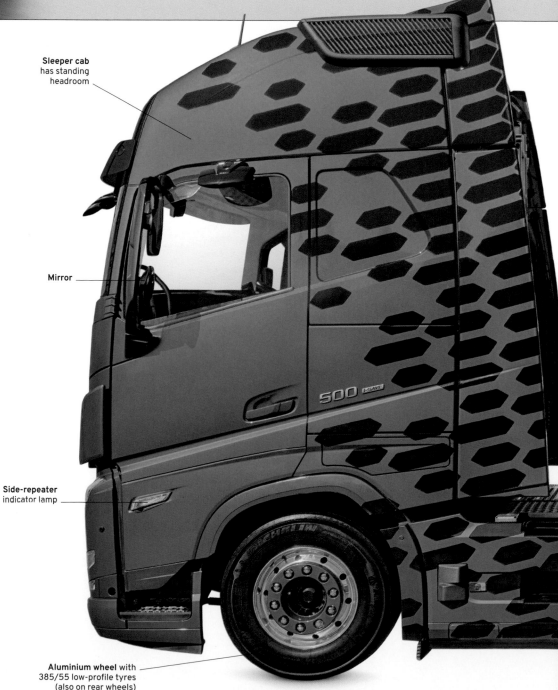

**Sleeper cab** has standing headroom

**Mirror**

**Side-repeater** indicator lamp

**Aluminium wheel** with 385/55 low-profile tyres (also on rear wheels)

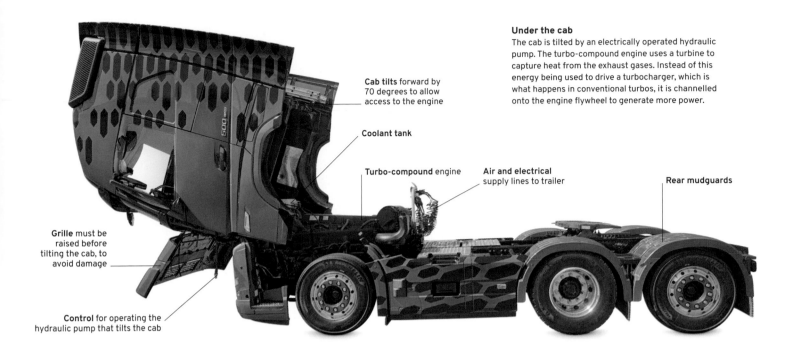

**Under the cab**
The cab is tilted by an electrically operated hydraulic pump. The turbo-compound engine uses a turbine to capture heat from the exhaust gases. Instead of this energy being used to drive a turbocharger, which is what happens in conventional turbos, it is channelled onto the engine flywheel to generate more power.

**Cab tilts** forward by 70 degrees to allow access to the engine

**Coolant tank**

**Turbo-compound** engine

**Air and electrical** supply lines to trailer

**Rear mudguards**

**Grille** must be raised before tilting the cab, to avoid damage

**Control** for operating the hydraulic pump that tilts the cab

**TAG-AXLE TRACTOR UNITS** form the bulk of tractor-trailer rigs. In most cases, the tag (an additional axle) is rigid and simply raised or lowered, depending on the weight being carried. This Volvo FH13, however, has a steering tag axle, which gives the truck greater manoeuvrability. The middle axle is the only one that is driven in this 6×2.

FH stands for forward-control high-entry. The spacious XL cab, with its high ceiling and plush leather trim, is designed to provide

superior levels of comfort for days or weeks on the road. It has a larger bed, increased standing room, extra luggage space in the overhead lockers, and a fridge – there is even an option for a microwave oven.

Computer-assisted technology helps to reduce fuel consumption: Volvo's I-Save reduces engine revs when on the open road; I-Shift promotes fast gear changes in automatic mode, maintaining smooth acceleration and deceleration when braking.

| SPECIFICATIONS | |
|---|---|
| Model | FH13-500TC |
| Origin | Sweden |
| Assembly | Sweden, Belgium |
| Production run | FH series from 1993–present |
| Weight | 7.7 tonnes (7.6 tons) |
| Payload | 44 tonnes (43 tons) |
| Engine | 6-cylinder diesel, up to 750 hp |
| Transmission | 12-speed automatic (manual override) |
| Maximum speed | Limited in Europe to 90 km/h (56 mph) |

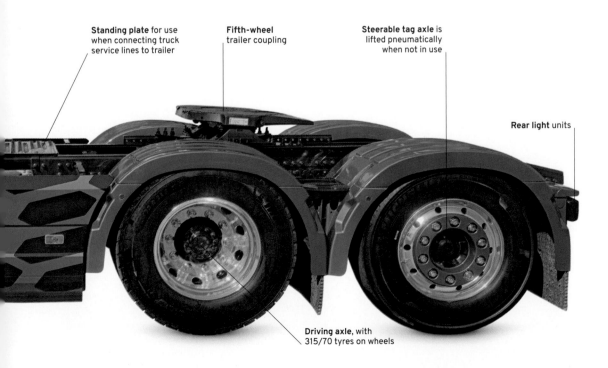

**Standing plate** for use when connecting truck service lines to trailer

**Fifth-wheel** trailer coupling

**Steerable tag axle** is lifted pneumatically when not in use

**Rear light** units

**Driving axle**, with 315/70 tyres on wheels

**Rolling inspiration**
Volvo was founded in 1926 by two employees at SKF bearing manufacturers, in Sweden. They chose the name Volvo (previously an SKF trademark) for the new vehicle company, which means "I roll" in Latin.

**Volvo FH13**
Although many trucks of this size use air suspension, the Volvo FH13 usually features coil springs. The rear bogie carrying the steerable tag axle, however, has full air suspension.

## THE EXTERIOR

The paintwork on this example has been custom-styled for maximum visual impact. Plenty of storage space is essential on long-haul trips, so the FH13 has four external lockers for stowing ratchet straps, spares, and personal items. Wind breakers behind the cab, which can be adjusted to suit the type of trailer being towed, increase aerodynamic efficiency and reduce fuel consumption. The FH13 sits very low, another factor – along with side skirts – that further helps to improve fuel efficiency. The Dura-Bright polished aluminium wheels enhance the vehicle's appearance.

**1.** Model badge  **2.** Cab door handle  **3.** Side-repeater indicator lamp  **4.** Cab step  **5.** Front Dura-Bright polished-aluminium wheel with body-coloured trim  **6.** Trailer service lines  **7.** No contact warning sign  **8.** Trailer de-coupling lever from fifth-wheel connection

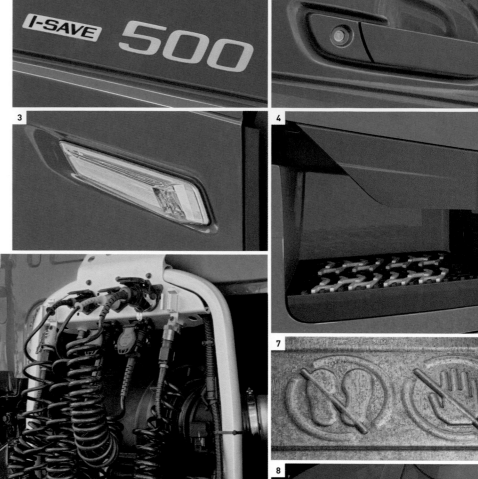

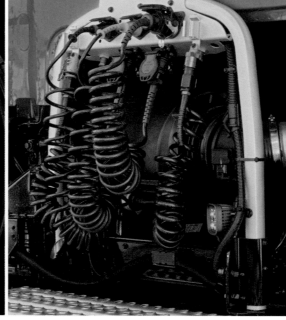

## THE ENGINE

The straight-six, turbocharged DT13T engine produces 500 hp and has a capacity of 12.88 litres. It has both an intercooler, which cools air as it passes through the engine, and a turbo-compound, which recovers energy and sends it to the flywheel, thus delivering more power without consuming more fuel. Peak power is delivered between 1,250–1,600 rpm, while maximum torque of 2,800 Nm is obtained between 900–1250 rpm. Volvo's I-Shift AT2812 is a 12-speed, fully automatic gearbox that has the option of full manual controls.

**9.** Access to radiator and fluid reservoirs   **10.** Six-cylinder, turbo-compound diesel engine

## THE INTERIOR

All the controls are angled towards the driver's seat. The instrument panel shows the engine gauges, and a sat-nav display is mounted to the left. An electric sunroof doubles as an escape hatch. The large interior allows for full standing height and extra storage lockers.

**11.** Cab layout   **12.** Electronic controls for the audio and air conditioning   **13.** Digital tachograph reader   **14.** Gear selector and brake   **15.** Control module for tag axle   **16.** Electric sunroof   **17.** Slide-out fridge   **18.** Driver's bed

# China's Robo Port

Modern trucks already have a degree of automation with aids such as active lane assist and active cruise control. These help to eliminate some common driving errors and improve safety. But autonomous (self-driving) trucks will take autonomy much further – they require little or even no human input.

## THE SHAPE OF THINGS TO COME

Some "autonomous" trucks will not be completely driverless, but some will only have human input remotely. While autonomous trucks have yet to get the go-ahead for full operation on public roads, they have already been used for a while in certain settings.

China has led the way in all kinds of driverless vehicles. In March 2023, Beijing granted permits for robotaxis with no driver or safety operator on board, but driverless trucks had been operating in certain Chinese ports for 18 months before that. Trucks used at Rizhao Port in the Shandong Province are Automated Guided Vehicles (AGVs) that use perception radar to spot and identify objects and hazards around the truck. This radar is effective in poor light and bad weather, and claims to increase safety and reduce accidents.

**In 2021, Rizhao Port in China** became the first fully automated container terminal in the world.

# The Future of Trucks

Huge changes are on the horizon for trucks. The diesel engine is destined to become a thing of the past as manufacturers look at alternative power sources. The same may one day apply to drivers, because technological advances are enabling sensors, artificial intelligence, and remote systems to control trucks without human involvement.

**Mercedes-Benz GenH2 prototype, 2020**
In September 2023, a prototype GenH2 travelled 1,047 km (650 miles) on a single fill of liquid hydrogen – with zero carbon dioxide emissions. This was an important milestone for non-fossil fuel trucks.

Until very recently, the internal combustion engine has been the driving force of trucks. While the fuels used have varied, and engine sizes and power have ranged greatly over the years, the internal combustion engine has remained a constant feature. But this is now changing, and, with climate change goals set to outlaw new diesel-engined trucks in the coming years, new, much-needed concepts for providing alternative motive power are coming to the fore.

Battery-electric trucks are already common in certain sectors, especially in urban distribution, where vans have to take short, multi-drop routes through towns and cities, often within low or zero-emission zones. Many manufacturers now offer electric versions of their diesel trucks, especially in Europe.

## Electric challenges
While electric trucks are proving a practical solution for companies that operate in areas local to their base, whether they will do the same for long-haul logistics is uncertain. Electric trucks have typically struggled to go much beyond a few hundred kilometres, due to the size and number of batteries required to power them, and the need to recharge. Nevertheless, technology is developing rapidly, and electric trucks are now becoming available that

their manufacturers claim can travel great distances on a single charge. Tesla's Semi, for example, can allegedly cover 800 km (500 miles), depending of the weight of the load carried. Other manufacturers are claiming that greater ranges will be available in the coming years.

## Power from hydrogen
Another power source currently being developed is hydrogen cells. Kenworth, Volvo, Hyundai, and DAF, among others, are already testing prototypes of hydrogen-powered trucks. These zero-exhaust-emission vehicles burn hydrogen to generate their own electricity on board, making them potentially viable for long-haul transport.

Using "green" hydrogen would be of most benefit to the environment. Green hydrogen is produced using energy from renewable sources, such as wind or solar. The production process for the alternative "blue" hydrogen uses natural gas, and so involves some carbon emissions.

Hydrogen-cell electric trucks are likely to be available commercially in the late 2020s. How truck stops will cater for the different needs of hydrogen, battery, and diesel vehicles remains to be seen.

**Westwell Qomolo "Q-Truck"**
The autonomous Q-Truck surveys its surroundings using LiDAR (light detection and ranging). An internal 3D map is generated, enabling precise movement and ensuring that collisions are avoided.

## Trucks without drivers
Autonomous (self-driving) trucks are already becoming a reality. In 2019, the Swedish manufacturer Einride, along with logistics company DB Schenker, put the first completely driverless truck (it does not even have a cab) onto a public road – albeit a short road between a warehouse and a terminal, reaching speeds typically below 32 km/h (20 mph).

In Thailand, Hutchinson Ports uses a fleet of autonomous "Q-Trucks" to move containers at its Laem Chabang terminal. Made by Westwell of China, Q-Trucks are equipped with a range

**Tesla Semi (3D illustration)**
The first Semis were delivered in 2022. Tesla claims that the three-motor Semi has triple the power of a typical diesel semi-truck. Tesla also has an electric pickup, called the Cybertruck.

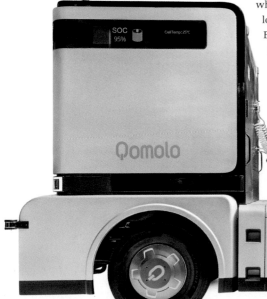

# Hydrogen fuel cell trucks, along with long-distance battery-electric trucks could be commonplace on roads by the end of the 2020s.

of sensors and AI machine-learning technology. A wireless charging system enables them to operate non-stop. Also involved in port work is Volvo's Vera, a low-slung, electric tractor unit with a "brain" housing the vehicle's technology where the cab would usually be. Vera began work in 2019, shifting goods from a logistics depot to a port terminal in Gothenburg, Sweden.

Scania's AXL concept truck, which also has no driver's cab, was launched in 2019 and is powered by biofuel. Designed for use in mines and construction sites, the AXL is steered and monitored by an "intelligent control environment", which means logistics software instructs the vehicle what to do.

It is in these largely off-road locations – mines, construction sites, freight yards and ports – that autonomous trucks are most likely to be used over the next few years.

While it is difficult to predict what trucks and trucking will be like in 2030 or 2050, the journey to the "truck of the future" will certainly be a fascinating and exciting one.

**TALKING POINT**

## Innovations in Trucking

A key recent innovation has been the development of advanced driver assistance systems (ADAS), which help to improve safety and reduce accidents and incidents of driver error. Tools such as active-lane-assist, which intervenes to keep a vehicle on course, are now standard on new trucks. In the next decade, ADAS will improve further. Advancements in software and hardware will make it possible to identify the speed and distance of objects ahead more quickly and accurately, making collision avoidance measures even more effective. Truck fleets will increasingly utilize remote technology, such as predictive maintenance. By analysing data from sensors and other sources, predictive maintenance can anticipate when parts are likely to break, malfunction, or need replacing – long before the driver knows that there is a problem.

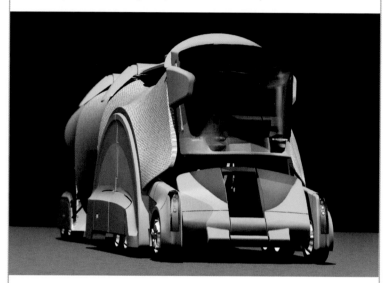

**Chameleon Truck concept by Haishan Deng** With a segmented, fabric-shelled body, it is designed to change its shape to fit its cargo, thus reducing inefficiency when not fully loaded.

# HOW TRUCKS WORK
# TRUCK
# TECHNOLOGY

# Engines

Whether fuelled by petrol or diesel, the engine is the beating heart of a truck. It must have the power required to haul heavy loads, but must also be reliable and economical to run and maintain. The internal combustion engine is now a mature technology, with over 150 years of continuous development behind it. Modern engines are now capable of running for 16,000–32,000 km (10,000–20,000 miles) between service intervals, and can typically be expected to cover many hundreds of thousands of kilometres before requiring a major overhaul. Today, petrol and diesel engines are gradually being replaced, with battery-electric and hydrogen fuel cells.

## ENGINE EVOLUTION

Engines first appeared in vehicles in the late 1800s. Over the following decades, their design, manufacture, and the materials used, were all improved to make them more powerful, reliable, and economical. More recently, turbocharging, which increases the air intake, and hybrid electrical assistance, combined with a greater understanding of combustion, have resulted in even greater torque, power, and efficiency, whilst also lowering emissions.

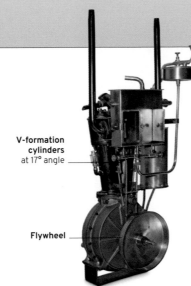

**V-formation cylinders** at 17° angle

**Flywheel**

**Two-cylinder Daimler engine**
In 1889, Gottlieb Daimler built the first V-engine, which produced 1.5 hp at 600 rpm. It was one of the early milestones in the development of the internal combustion engine for vehicles.

**Four cylinders** behind the valve springs

**Foot pedals** for clutch, accelerator, and brakes

**Manual starter handle**

**1911 Four-cylinder engine**
Having four combustion cylinders arranged in a line, or horizontally opposed, produces more torque (twisting power) than in two-cylinder engines.

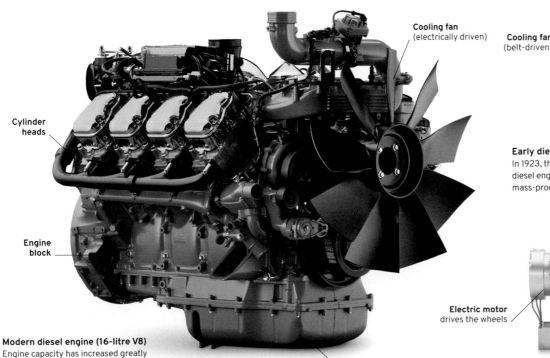

**Cylinder heads**

**Engine block**

**Cooling fan** (electrically driven)

**Modern diesel engine (16-litre V8)**
Engine capacity has increased greatly over the years. A typical 16-litre V8 engine is now capable of producing more than 670 hp and over 3,000 Nm of torque.

**Oil pan**

**Valves** (intake and exhaust)

**Cooling fan** (belt-driven)

**Early diesel engine**
In 1923, the four-cylinder Benz OB 2 precombustion diesel engine was introduced. It became the world's first mass-produced diesel engine for commercial vehicles.

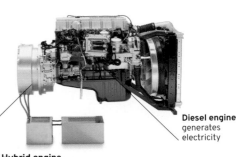

**Electric motor** drives the wheels

**Diesel engine** generates electricity

**Hybrid engine**
A hybrid combines a diesel engine with an electric motor/generator unit. The engine and regenerative braking recharge the battery that powers the motor.

## ENGINE ACCESS

To ensure reliability and to keep trucks working at peak efficiency, engines need regular maintenance. This includes changing oil, water, filters for air and oil, and worn components such as fan belts, spark plugs (in petrol engines) and glow plugs (in diesel engines). Ease of access to these components is essential. The earliest approach was to design trucks with engine covers, or bonnets, that can be lifted. Today, cab-over trucks are equipped with a tilting cab: the entire cab lifts forwards, giving clear, unrestricted access to the engine's key maintenance points.

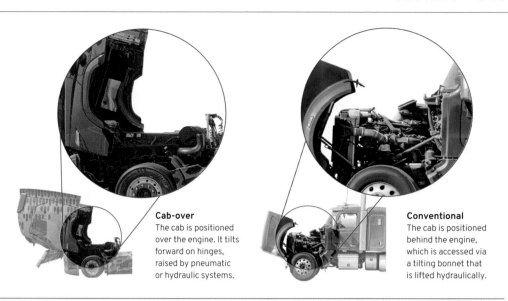

**Cab-over**
The cab is positioned over the engine. It tilts forward on hinges, raised by pneumatic or hydraulic systems.

**Conventional**
The cab is positioned behind the engine, which is accessed via a tilting bonnet that is lifted hydraulically.

## ELECTRIC TRUCK

Due to a combination of environmental, economic, and technological factors, truck manufacturers are moving away from internal combustion engines, and trucks powered by petrol or diesel are being phased out. Manufacturers are now developing battery-electric vehicles that produce no carbon emissions from their propulsion systems.

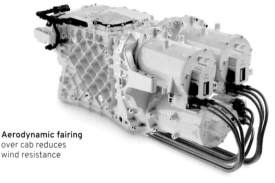

**Electric motor**
Battery-electric trucks use large lithium ion batteries to power motors that propel the truck. The advantage of electric motors is that they are lighter and more reliable than conventional engines. They also produce maximum torque from zero revs, which makes them ideal for haulage. However, the batteries must be recharged, which limits the range of these trucks. The batteries are also heavy, which reduces a truck's load capacity.

**Aerodynamic fairing**
over cab reduces wind resistance

**Inverter** converts DC electric current to AC

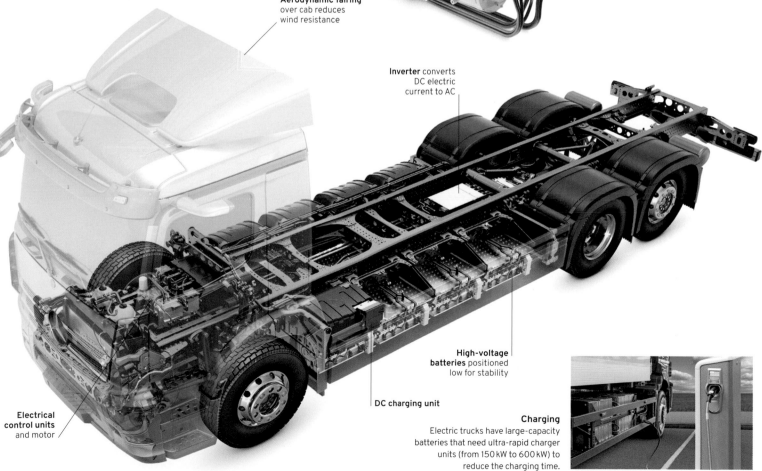

**Electrical control units** and motor

**DC charging unit**

**High-voltage batteries** positioned low for stability

**Charging**
Electric trucks have large-capacity batteries that need ultra-rapid charger units (from 150 kW to 600 kW) to reduce the charging time.

# Chassis and Suspension

The role of a truck's chassis is to support the weight of the truck and its payload. Typically, the materials used to construct vehicle chassis and frames include carbon steel for strength, and aluminium alloys to achieve a more lightweight construction. The suspension connects the chassis to the wheels, and is optimized to provide predictable handling and stability over different surfaces. The challenge is to provide as smooth a ride as possible under a wide range of operating conditions, both while laden and unladen. A rigid chassis is essential for a truck, as any flexing of the frame would compromise the handling and stability of the vehicle.

## CHASSIS AND WHEEL CONFIGURATION

Trucks with different wheel configurations are able to haul different loads. The number of wheels on each axle and the number of axles driven by the engine vary accordingly. The heavier the load, the greater the number of wheels needed to distribute the weight and prevent overloading the tyres or road surface.

**4×2** Four wheels; only the rear axle is driven.

**4×4** Four wheels; rear and front axle are both driven.

**6×2** Six wheels; only the second axle is driven.

**6×4** Six wheels; both rear axles are driven.

**6×6** Six wheels; all three axles are driven.

**8×2** Eight wheels; the leading rear axle is driven, two front axles steer.

## TRACTION

As the load increases, a truck needs to exert more friction on the road in order to have the traction to pull it. Additional axles that can be retracted when running without a load can have a number of different positions.

**Driven tag axle**
An additional driven axle behind the leading driven axle improves traction.

**Undriven tag axle**
An extra undriven axle behind the driven axle gives better load bearing.

**Mid-lift**
An undriven axle behind the leading axle gives support and stability.

**Mini mid-lift**
Smaller diameter wheels and tyres allow increased overall load weight.

### TECHNOLOGY

## Fifth Wheel

The basis of the semi-articulated truck, the fifth wheel provides an articulated joint between the tractor and trailer. The flat, horseshoe-shaped plate with a locking mechanism allows a trailer to be quickly and securely connected and disconnected. The kingpin from the trailer locks into place in the fifth wheel and can pivot within the connection to avoid tipping when the truck turns.

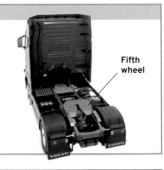

**Fifth wheel**

## SUSPENSION

The primary function of suspension is to absorb the movement caused by an uneven road surface, and thus maintain stability. Suspension also spreads the load evenly across all the axles and ensures that the wheels stay in contact with the road to provide traction. Steel leaf or steel coil springs absorb the force, while hydraulic dampers control the rebound. The vast majority of trucks today use air springs, which can be adjusted according to the weight of the load.

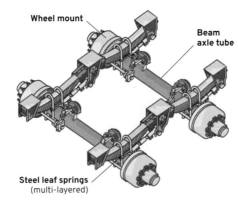

**Wheel mount**

**Beam axle tube**

**Steel leaf springs** (multi-layered)

**Leaf-spring suspension**
This type of mechanical suspension consists of multiple curved and flexible strips of steel. It has changed little since being used on horse-drawn carts.

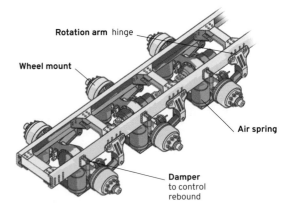

**Rotation arm** hinge

**Wheel mount**

**Air spring**

**Damper** to control rebound

**Air suspension**
The vehicle's weight is supported by compressed air in airbags. Air suspension allows fine tuning of the ride height by increasing or lowering the pressure.

# Inside the Cab

The cab of a truck is the driver's cockpit, workplace, office, and (if the cab is equipped with a bunk) their home from home. It needs to be ergonomically designed do make operating the vehicle as easy and safe as possible. The cab has its own suspension system to absorb shocks and vibrations from the road surface and uneven terrain. This helps to prevent excessive wear and tear on the cab and its components. It also ensures that the occupants in the cab experience a smooth and comfortable ride, even when the truck encounters bumps and potholes, or rough roads. A well-designed cab will reduce fatigue for drivers who have to spend days on the road at a time.

## FIXTURES/FITTINGS

As well as being equipped with the standard driver controls for steering, accelerator, brakes, and transmission, the cab is kitted out with interior lighting, ventilation controls, and communication and entertainment systems. The interior of a modern truck cab also has fully adjustable seats, and stowage compartments for work documents, clothing, and the driver's personal effects.

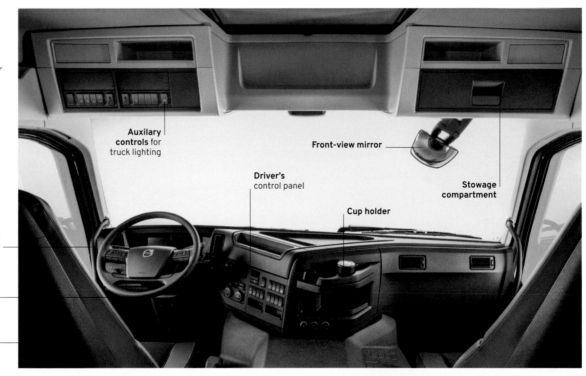

Auxilary **controls** for truck lighting

Front-view mirror

Driver's control panel

Stowage compartment

Cup holder

**Grab handle** to aid access

**Steering wheel** is adjustable for reach and height

**Driver's seat** is fully adjustable and suspended

## DRIVER AIDS

Modern trucks are equipped with a myriad of driver aids. GPS (global positioning system) aids navigation and provides traffic information, while ABS (anti-lock braking system), automated transmission, adaptive cruise control, lane-control departure warnings and assistance, and a collision-avoidance system all help to reduce the strain on the driver. Sensors and cameras provide blind-spot alerts, while improved all-round visibility and trailer-stability control systems allow safer towing. A special monitor alerts a driver who shows signs of inattention or fatigue.

**Digital tachograph/electronic logbook**
This is an electronic device that records and monitors the driver's working hours, driving time, and rest breaks.

**Mirror camera**
Many trucks now employ cameras and screens to monitor traffic conditions around the vehicle, rather than glass mirrors.

# Glossary

**4×4**
Shorthand for a four-wheeled vehicle in which the engine power goes to all four wheels. In a 4×2 vehicle, only two wheels are driven. *See also* four-wheel drive.

**6x2**
Shorthand for a six-wheeled vehicle in which the engine power goes to just two wheels.

**advanced driver-assistance systems (ADAS)**
Electronic systems that assist drivers with the safe operation of a vehicle. Technologies come in many forms and are primarily focused on avoiding collisions.

**aerodynamic**
Having a streamlined shape, with smooth contours that reduce air resistance when in motion. The more aerodynamic a vehicle's shape, the less fuel the vehicle uses.

**air suspension**
A type of vehicle suspension that supports its load on air-filled rubber bags as opposed to conventional steel springs. Powered by an electric or engine-driven air pump or compressor, it is used in heavy vehicles to improve driver comfort and reduce wear and tear on components.

**A-train**
A truck-trailer combination that has a basic dolly, or platform on wheels, hitched up to the rear of the forward trailer with a hook.

**automated guided vehicle (AGV)**
A driverless vehicle that uses advanced technologies, such as radars, cameras, and sensors to operate. *See* autonomous vehicle.

**amphibious vehicle**
A vehicle that can travel on water as well as drive on land.

**articulated**
A vehicle that is made up of two or more sections connected by a pivoting joint, making the vehicle more flexible and allowing it to make sharper turns.

**autonomous vehicle**
A vehicle that uses technologies such as cameras and artificial intelligence (AI) to operate without human intervention. Some autonomous trucks are already in use around the world. *See also* automated guided vehicle (AGV).

**axle**
A rod or spindle that goes through the centre of the wheels. The axle is either fixed to the wheels and rotates with them, or is fixed to the vehicle and stays still as the wheels rotate around the axle.

**biodiesel**
A cleaner, less environmentally damaging alternative fuel for diesel engines, made from natural and renewable sources, including vegetable oils and animal fats.

**bobtail**
The nickname given to a tractor unit with no trailer attached to it. It is what a truck is called after dropping off a trailer and before it has picked up the next one. Driving a bobtail truck, referred to as bobtailing, poses risks, because a truck with no trailer can be difficult to manoeuvre.

**bonnet**
A hinged covering for a truck's engine.

**B-train**
Two semi-trailers linked together and pulled by a truck. The leading semi-trailer is attached to the truck by a fifth-wheel coupling. An additional fifth-wheel coupling at the rear of this semi is used to attach the second trailer.

**cab**
Sometimes referred to as a cabin, the cab is the enclosed section of a truck in which the driver sits and operates the controls.

**cab-over**
Also known as a cab-over-engine (COE), the cab-over is a predominantly European truck style with a vertical front, or flatter nose, and no bonnet. The cab sits over the engine.

**cargo**
The goods, wares, merchandise, or materials carried from one place to another by a truck.

**cc (cubic centimetres)**
The standard volumetric measurement of cylinder capacity – and therefore engine size – for engines in Europe and Japan.

**chassis**
A load-bearing frame on wheels that carries the mechanical parts to which the truck body is attached. Most trucks that exceed 3.5 tonnes (3.4 tons) have a separate chassis.

**commercial vehicle (CV)**
Any motor vehicle that transports goods or materials rather than passengers.

**compressed natural gas (CNG)**
A less environmentally damaging fuel alternative to petrol and diesel used in heavy-goods vehicles. It is made by compressing natural gas (primarily methane) to less than 1 per cent of its volume at normal atmospheric pressure.

**conventional**
A truck where the cab is positioned behind the engine compartment, much like in a typical car.

**cu in (cubic inches)**
A volumetric measurement of cylinder capacity – and therefore engine size – for engines in the US (1 cu in is approximately 16.4 cc). It was replaced by the litre from the 1970s onwards.

**cylinder**
An enclosed chamber in which a piston moves to produce motive power that is transmitted to wheels. In a steam engine, the piston is made to move by the force of high-pressure steam acting against it (an external combustion engine). An internal combustion engine produces its power by burning liquid fuel that is introduced into the cylinder(s).

**cylinder block**
The body, usually of cast metal, into which cylinders are bored to carry the pistons in an internal combustion engine, and to which the cylinder head or heads attach.

**cylinder head**
The upper part of an engine, attached to the top of the cylinder block. It contains components such as the spark plugs that ignite the fuel in the cylinders, and the exhaust valves.

**day cab**
A type of truck cab that does not have a bed for the driver. Lighter than their sleeper cab counterparts, day cabs are used on trucks that make short-haul deliveries.

**dekotora**
Short for "decorated truck", dekotora is the Japanese practice of adding vivid embellishments to a truck, such as LED lights, chrome fixtures, and detailed, highly colourful paintwork.

**diesel**
A fuel distilled from crude oil comprising 75 per cent saturated hydrocarbons and 25 per cent aromatic hydrocarbons. Diesel is supplied as either white diesel (derv) for road use, or red diesel (gas oil) for "off-highway" applications, such as for agricultural use in tractors. Standard refined diesel, with no additives, starts to "gel" at −3°C (27.6°F), meaning that the waxes it contains begin to crystallize. It freezes at −8°C (17.6°F).

**drayage**
A short-haul trip carrying freight from one place to another, often part of a longer overall route. Drayage trucks are usually heavy-duty diesel trucks.

**drivetrain**
The group of components, including transmission and driveshafts, that delivers power from the engine to the wheels.

**electric vehicle (EV)**
A vehicle powered by a battery that can be charged using an electric charging point.

**exhaust pipe**
Also known as a tailpipe, the pipe attached to a vehicle through which engine exhaust gases are expelled.

**fifth wheel**
A horseshoe-shaped coupling device that provides the link between the trailer and the truck. The kingpin from the trailer connects to the truck's fifth wheel.

**flatbed**
A truck with a flat, level area at the back with no roof or sides. Flatbeds are used to transport heavy objects.

**fleet**
A group of vehicles owned and operated by an organization or company, rather than by an individual.

**four-stroke engine**
A type of internal combustion engine used in many trucks. It operates through four distinct phases: intake, compression, power, and exhaust, and delivers one power stroke for every two revolutions of the crankshaft. *See also* two-stroke engine.

**four-wheel drive (4WD)**
Power transmitted to all four wheels, instead of two; 4WD gives better traction under difficult conditions, such as muddy terrain.

**freight**
The transportation of goods, especially by a commercial carrier. Unlike cargo, which refers specifically to the goods being transported, freight encompasses the entire logistics process, from pickup to delivery.

**front-wheel drive**
Power transmitted to the two front wheels of a vehicle only. This lightens the vehicle, because drive shafts are not required to send power to the rear wheels.

**gas turbine**
A jet-type rotary engine that draws its energy from the continuous burning of a flow of fuel-air mixture, which drives a turbine. It was used experimentally in vehicles, but it proved too slow-reacting to directly replace the piston engine.

**gear**
A toothed or cogged machine part that meshes and rotates with other such parts to transmit power and torque. Shifting between gears alters the rate at which energy is changed into motion.

**grille**
A screen or grating covering an opening yet enabling ventilation. Truck grilles typically cover the radiator and engine.

**gross vehicle weight (GVW)**
The maximum permissible weight of a vehicle, including the load (driver, passengers, goods, etc.) that can be carried safely while on the road.

**haulage**
The commercial transport of goods by road between manufacturers, suppliers, and consumers.

**heavy truck**
Generally this refers to trucks with GVW (gross vehicle weight) ratings that exceed 7.5 tonnes (7.4 tons). The exact definition varies around the world.

**horsepower (hp)**
A unit of power used to measure the output of an engine that is equal to the power required to move 75 kg one metre (550 lb one foot) in one second. The term was adopted in the 18th century by British engineer James Watt to compare the output of steam engines with the amount of work performed by a single draft horse.

**hybrid**
A vehicle-propulsion technology that combines both electric and petrol/diesel power. In electric-power mode, emissions are reduced during urban driving, while the fossil-fuel mode gives sustained power for motorway cruising, and recharges the battery at the same time.

**hydraulics**
The use of a fluid medium under pressure in a cylinder to power and control actions, such as lifting and lowering the arm of a crane or tailgate lift. Hydraulics are also used for brakes on trucks.

**hydrogen fuel**
A "clean" fuel that can be used to power vehicle engines. When burned, hydrogen emits water vapour and warm air.

**hydrogen fuel cell**
A device that converts the chemical energy of hydrogen into electricity, using methods such as natural gas reforming or electrolysis. Vehicles that use this technology are called "fuel cell electric vehicles" (FCEV).

**hydrotreated vegetable oil (HVO)**
Also known as renewable diesel, HVO is a biofuel made by hydrogenating vegetable oils and plant waste.

**jingle truck**
An elaborately decorated truck from Pakistan. The name comes from the sound of the bells festooning the exterior of such trucks.

**kingpin**
The connector fitted to the trailer that engages with the fifth wheel coupling on the tractor unit. It is the main pivot in the steering mechanism of the coupled rig.

**layover**
The term for when a trucker is forced to delay their journey for an extended period of time.

**leaf spring**
A basic form of suspension, also known as a "cart spring". The spring comprises overlaid arcs (or leaves) of steel that are fixed to the underside of a vehicle, forming a shock-absorbing cushion onto which the axle presses.

**liftgate**
A hydraulic platform at the rear of a truck for loading or unloading heavy goods.

**light truck**
Generally, trucks with GVW (gross vehicle weight) ratings that do not exceed 4.5 tonnes (4.4 tons). Utility and sport-utility vehicles are classified as light trucks even when used as family cars. *See also* heavy truck.

**liquefied natural gas (LNG)**
Natural gas that has been cooled to a liquid state. LNG is most often used to power large, long-distance trucks.

**long combination vehicle (LCV)**
Any combination of a truck and two or more semi-trailers. LCVs offer an effective and fuel-efficient way to transport heavy loads over long distances.

**long-haul**
The transportation of goods over distances of 400 km (250 miles) or more.

**monster truck**
A pickup truck with oversized tyres and modified suspension, typically used for racing over rough terrain and performing stunts in spectacular truck shows.

**nominal horsepower (nhp)**
A very approximate measure of the power of early steam engines, which was used by manufacturers to indicate an engine's power when compared to that of a horse.

**panel van**
A small van without rear or side windows, or rear passenger seats. Panel vans are sometimes referred to as delivery vans.

**payload**
The amount of weight a truck or vehicle can safely carry, which does not include the weight of the truck.

**petrol**
Processed crude oil (petroleum) that is used as a fuel in internal combustion engines.

**plug-in hybrid (PHEV)**
A hybrid battery-electric vehicle that still has an internal combustion engine. It can be plugged in to charge or charged by the engine as it is driven.

**pickup**
A truck with an enclosed cabin for the driver and passenger at the front, and an open cargo bed with low sides at the rear for transporting materials.

**prime mover**
A heavy-duty truck, also known as a "puller vehicle", that is used to tow a semi-trailer. Prime movers have excellent motive power and are capable of carrying heavy loads.

**rampside**
A swing-down panel on the passenger side of a pickup that allows heavy cargo to be easily loaded into its bed.

**rating (horsepower)**
The horsepower rating of an engine can be given in many different ways. The ratings can be shown as brake horsepower (bhp); draw-bar horsepower (dhp); metric horsepower (din); Society of Automotive Engineers (SAE) specification; or in kilowatts (kW). Each rating has its own specific method of measuring criteria.

**road train**
A line of linked trailers pulled by a single truck. Road trains are more efficient than using multiple semi-trailer trucks for long-distance haulage. *See also* LCV.

**semi-trailer**
A large trailer with wheels at the back. The trailer is attached to a tractor unit that supports it and pulls it along.

**sleeper cab**
A cab with a bunk bed behind the seating area, in which a driver can rest or sleep during breaks on long journeys.

**steam engine**
An engine that uses high-pressure steam from boiling water to drive the pistons in the cylinders.

**susies**
The electrical lines that provide a connection between a truck and a trailer.

**suspension**
A system that cushions a vehicle's structure (and occupants) from the motion of the wheels as they travel over uneven road surfaces, giving a smoother ride.

**tachograph**
An instrument that records information about the speed and distance a truck driver has driven, as well as the length of time they have been on the road. An essential piece of safety equipment, it ensures that a driver takes regular breaks and has rest days. Since May 2006, all new commercial vehicles in the EU must be fitted with digital tachographs. In the US, truck drivers have been required to record their working hours electronically since December 2017.

**tailgate**
The hinged opening at the back of a truck that can be lowered to load goods.

**traction engine**
A self-propelled engine driven by steam. The engine could be used for towing and/or powering machinery. They were also used for general haulage. Traction engines were introduced in the mid-19th century. Though now outdated, they continue to be used.

**tractor unit**
Tractor units are used to pull one or more trailers. They do not have a load area and do not carry cargo directly.

**torque**
A measurement of the maximum twisting force an engine can generate. The more torque, the greater an engine's capacity to perform work.

**transmission**
Part of the drivetrain, the transmission uses gears to ensure that the right amount of power goes to the wheels to drive at a given speed. The term is sometimes used to refer to the gearbox alone.

**truck stop**
A large roadside service station, usually with a restaurant, where drivers can stop to rest and buy fuel, food, and supplies.

**turbocharger**
A device fitted between an engine's inlet and exhaust systems that uses the exhaust gases to drive a turbine. This, in turn, drives a compressor that forces air into the inlet, giving the engine more power.

**two-stroke engine**
An engine that gives one power stroke for every revolution of the crankshaft.

**tyre tread**
The pattern of moulded rubber on the surface of a tyre. It is non-reversible and designed to be self-cleaning in one direction only. Tyre treads grip the road, enabling the wheels to convert the engine's power into motion.

**V4, V6, V8, V10, V12, V16**
The designations for engines designed with their cylinders arranged in a V-formation for compactness. The numbers relate to the number of cylinders in each engine.

**wheelbase**
The distance between the centres of the front and rear axles. On trucks with more than two axles, it is measured as the distance between the steering axle and the centre of the driving axle group.

**wrecker**
A recovery vehicle, used to transport stranded vehicles.

# Index

All page references are given in *italics*.

# Acknowledgments

**Dorling Kindersley** would like to extend thanks to the following people for their help with making the book:

Nic Dean and Steve Behan for picture research; Steve Crozier for retouching; Phil Gamble for illustration; Shipra Jain and Chhaya Sajwan for design assistance; Janashree Singha and Abhijit Dutta for editorial assistance; Syed Md Farhan and Ashok Kumar for DTP assistance; Sonia Charbonnier for translation help; Alicia Williamson, Nick Funnell, Steve Setford, and John Andrews for editing; Belinda Gallagher for proofreading; and Nic Nicolas for indexing.

**The publisher would like to thank** the following manufacturers for their help with making the book:

**DAF**
Hugo van der Goeslaan 1, 5643 TW Eindhoven, The Netherlands
www.daf.com

**Ford Motor Company**
1 American Rd, Dearborn, MI 48126, United States
www.ford.com

**GMC**
100 Renaissance Center, Detroit, MI 48243, United States
www.gmc.com

**Iveco**
Iveco S.P.A., Via Puglia 35, 10156 Torino, Italy
www.ivecogroup.com

**Scania**
Scania AB (publ), SE-151 87 Södertälje, Sweden
www.scania.com

**MAN**
MAN Truck & Bus SE, Dachauer Straße 667, 80995, Munich
www.man.eu

**Nissan Motor Co., Ltd.**
1-1, Takashima 1-chome, Nishi-ku, Yokohama-shi, Kanagawa 220-8686, Japan
www.nissan-global.com

**Mercedes Benz Group**
Mercedesstraße 120, 70372, Stuttgart-Untertürkheim, Germany
https://group.mercedes-benz.com

**Renault Group**
122-122 bis avenue du Général Leclerc, 92100, Boulogne-Billancourt, France
www.renaultgroup.com

**Toyota**
1 Toyota-Cho, Toyota City, Aichi Prefecture 471-8571, Japan
https://global.toyota/

**AB Volvo**
AB Volvo, SE-405 08 Gothenburg, Sweden
www.volvogroup.com

The publisher would also like to thank the following museums and companies for giving access to their trucks for photography and for their help with picture research:

**American Truck Historical Society**
10380 N. Ambassador Drive, #101, Kansas City, Missouri 64153, United States
https://aths.org/

**Atkinson Vos**
Wenning Avenue, Bentham, Lancaster, LA2 7LW
www.unimogs.co.uk

**Berliet Foundation**
39, avenue Esquirol, 69003 Lyon, France
www.fondationberliet.org

**Bonhams**
101 New Bond St, London, W1S 1SR, UK
www.bonhams.com

**British Commercial Vehicle Museum**
King Street, Leyland, Lancashire, PR25 2LE, UK
www.britishcommercialvehiclemuseum.com

**British Motor Museum**
Banbury Road, Gaydon, Warwickshire, CV35 0BJ
www.britishmotormuseum.co.uk

**Devon Truck Show**
www.devontruckshow.co.uk

**Dover Transport Museum**
Willingdon Road, Whitfield, Dover, CT16 2JX
www.dovertransportmuseum.org.uk

**Grundon Waste Management**
Thames House, Oxford Road, Benson, OX10 6LX
www.grundon.com

**Hyman Ltd**
2310 Chaffee Drive, St. Louis, MO 63146, United States
https://hymanltd.com/

**Louwman Museum**
Leidsestraatweg 5, 2594 BB Den Haag, The Netherlands
www.louwmanmuseum.nl

**Mecum Auctions**
445 S. Main Street, Walworth, WI 53184, United States
www.mecum.com

**Sotheby's**
1334 York Avenue, New York, New York 10021, United States
www.sothebys.com

**Volvo Group UK Ltd**
Wedgnock Lane, Warwick, CV34 5YA
www.volvogroup.com

The publisher would like to thank the following for their kind permission to reproduce their photographs:

Key: a-above; b-below/bottom; c-centre; f-far; l-left; r-right; t-top

**2-3 From the collections of the American Truck Historical Society:** (c). **9 Hyman Ltd.** **www.hymanltd.com:** (bl). **12 The Advertising Archives.** **13 akg-images:** (cra). **Getty Images:** Keystone-France/Gamma-Keystone (clb). **14 Dorling Kindersley:** Milestones Museum (cla). **Fondation de l'Automobile Marius Berliet, Lyon, France:** (br). **Louwman Museum, The Hague:** (bl). **15 Dorling Kindersley:** Milestones Museum (cl). **16 Louwman Museum, The Hague:** (tr). **Mercedes-Benz Classic:** (cl, bl). **Wikipedia:** (br). **17 Bonhams:** (tr). **Fondation de l'Automobile Marius Berliet, Lyon, France:** (clb). **Gary Alan Nelson:** (bc). **Hyman Ltd.** **www.hymanltd.com:** (crb). **18 Courtesy of BCVM LEYLAND:** (tl). **22-23 Dorling Kindersley:** Milestones Museum (tc). **22 Bonhams:** (bl). **Fondation de l'Automobile Marius Berliet, Lyon, France:** (cl). **Gary Alan Nelson:** (br). **Hyman Ltd.** **www.hymanltd.com:** (cr). **23 Dorling Kindersley:** Milestones Museum (cl). **Gary Alan Nelson:** (bl, br, tr). **24-25 Mary Evans Picture Library:** The March of the Women Collection. **26 Alamy Stock Photo:** Barry Gore (cla). **Courtesy of BCVM LEYLAND:** (tl, cr). **27 Alamy Stock Photo:** NPC Collection (b). **DAF Trucks N.V.:** (tr). **Courtesy of BCVM LEYLAND:** (tl). **28 Alamy Stock Photo:** pbpgalleries (bl). **Bonhams:** (br). **Hyman Ltd. www.hymanltd.com:** (cb). **Louwman Museum, The Hague:** (tl). **28-29 Louwman Museum, The Hague:** (tc). **29 Bonhams:** (bl). **Louwman Museum, The Hague:** (tr). **Courtesy of RM Sotheby's:** (br). **30 Alamy Stock Photo:** piemags/NSC (tl). **34 Copyright © Ford Motor Company. All rights reserved.:** (bl). **Getty Images:** Hulton-Deutsch Collection/Corbis (tl); Interim Archives (cl). **34-35 Copyright © Ford Motor Company. All rights reserved.:** (bc). **35 Copyright © Ford Motor Company. All rights reserved.:** (cr, ftr, tr, br). **36 Fondation de l'Automobile Marius Berliet, Lyon, France:** (cl, bl). **Snapshooter46:** (tr). **37 Alamy Stock Photo:** Jacques Brinon/Associated Press (cl). **Bonhams:** (tr). **Dorling Kindersley:** Milestones Museum (b). **The Packard's restoration project was completed at Dave Lockard's home in York Springs, Pennsylvania by many volunteers and now resides at America's Packard Museum, Dayton, Ohio.:** (cr). **38-39 Scania CV AB:** (c). **40 Alamy Stock Photo:** Science History Images/Photo Researchers (bl). **Getty Images:** Topical Press Agency/Hulton Archive (t). **41 Alamy Stock Photo:** National Motor Museum/Heritage Images (b); ZarkePix (cra). **42-43 Hyman Ltd. www.hymanltd.com:** (c). **44 akg-images:** Heritage-Images/National Motor Museum. **45 Getty Images:** Hulton Archive (ca). **Mary Evans Picture Library:** Sueddeutsche Zeitung Photo (crb). **46 Fondation de l'Automobile Marius Berliet, Lyon, France:** (cl). **Gary Alan Nelson:** (tc, crb). **Getty Images:** Fox Photos (bl). **47 Dorling Kindersley:** Keystone Tractor Works (t); S M Sheppard (br). **Fondation de l'Automobile Marius Berliet, Lyon, France:** (bl). **Gary Alan Nelson:** (cl). **48 Fondation de l'Automobile Marius Berliet, Lyon, France:** (tl, cra, bl). **49 Alamy Stock Photo:** Taina Sohlman (tl). **Fondation de l'Automobile Marius Berliet, Lyon, France:** (ftl, tl, tr, cra). **© Renault Trucks SAS:** (ftr). **50 orangevolvobusdriver4u:** (cl). **Thomas Tutchek:** (bl). **51 Dorling Kindersley:** Milestones Museum (tl, bl, br). **52 Alamy Stock Photo:** Mark Phillips (cra). **Copyright © Ford Motor Company. All rights reserved.:** (tl). **56 From the collections of the American Truck Historical Society:** (tl, br). **Old International Trucks:** (bl). **57 Dorling Kindersley:** Keystone Tractor Works (tr). **From the collections of the American Truck Historical Society:** (br). **Gary Alan Nelson:** (tl, bl). **58 MAN Truck & Bus Historical Archive:** (tl, c, bc). **59 MAN Truck & Bus Historical Archive:** (ftl, tl, cr, bl). **60-61 Getty Images:** Fox Photos/Hulton Archive. **62 EmmeBi Photos:** (br). **Hyman Ltd. www.hymanltd.com:** (cl, bl).

**Wikipedia:** Dmitry Ivanov (cr). **63 Fondation de l'Automobile Marius Berliet, Lyon, France:** (bl). **Hyman Ltd. www.hymanltd.com:** (cl). **Nissan Heritage Collection:** (br, cr). **64 Courtesy of BCVM LEYLAND:** (tl). **68 Dorling Kindersley:** Milestones Museum (cla). **68-69 Collection Ghezzi/Luchin Tn:** (bc). **69 Alamy Stock Photo:** Grzegorz Czapski (br). **EmmeBi Photos:** (tr). **Gary Alan Nelson:** (cla). **Mercedes-Benz Classic:** (tl). **70-71 Getty Images:** Hirz. **72 Bonhams:** (tr). **Gary Alan Nelson:** (clb). **Hyman Ltd. www. hymanltd.com:** (c). **Perico001:** (bl). **72-73 Wikipedia:** TexacoTanker/ Transport World (bc). **73 Gary Alan Nelson:** (cla). **Getty Images/iStock:** Gwengoat (tr). **Hyman Ltd. www. hymanltd.com:** (tl, cr). **74 Alamy Stock Photo:** Associated Press (bl). **74-75 Alamy Stock Photo:** Paul Mayall Australia (br). **75 Dreamstime. com:** Welcomia (tr). **76 Dorling Kindersley:** Keystone Tractor Works (bl, br). **77 Alamy Stock Photo:** Serguei Dratchev (clb). **Dorling Kindersley:** Keystone Tractor Works (tl). **Photo courtesy of Mecum Auctions, Inc.:** (cr). **Volvo Car Group:** (tr). **78-79 Hyman Ltd. www. hymanltd.com:** (c). **78 Doylestown Fire Company No. 1:** (tl). **Hyman Ltd. www.hymanltd.com:** (bl, bc, br). **79 Hyman Ltd. www.hymanltd.com:** (tl, tc, tr, ca, cr). **82 Getty Images:** National Motor Museum/Heritage Images. **83 Copyright © Ford Motor Company. All rights reserved.:** (cr). **Getty Images:** Hulton Archive (bl). **Mary Evans Picture Library:** Interfoto/Pulfer (tc). **84 B.C. Vintage Truck Museum:** (bl). **85 Alamy Stock Photo: Steven Bennett** (tl). **86 US Army:** (tl). **90-91 Mary Evans Picture Library:** Robert Hunt Collection. **92 Volvo Trucks:** (tl, bl, br, cr). **93 Volvo Trucks:** (ftl, bc). **94 Hyman Ltd. www.hymanltd.com:** (cl, bl, tr). **95 Mercedes-Benz Classic:** (cl). **Toyota Automobile Museum:** (tr, cr). **96 Hyman Ltd. www.hymanltd.com:** (cl, bl, bc). **SuperStock: Huntington Library** (tl). **96-97 Hyman Ltd. www. hymanltd.com:** (c). **97 Hyman Ltd. www.hymanltd.com:** (tc, tr, ca, cra, cr, br). **98 Alamy Stock Photo:** Viktor Karasev (cla); Oldtimer (crb); Wahavi (bl). **Koskin Import:** (br). **100 Jan Barnier:** (tr). **Dorling Kindersley:** Keystone Tractor Works (cl, b). **101 Alamy Stock Photo:** Bernard Menigault (cl). **Dorling Kindersley:** Keystone Tractor Works (tl, tr, br, crb). **102 Alamy Stock Photo:** Sue Thatcher (crb). **Fondation de l'Automobile Marius Berliet, Lyon, France:** (clb, br). **Monde Auto Photo Passion/Flickr:** (tr). **Courtesy of RM Sotheby's:** (cra). **103 Thomas Bersy:** (tr). **Fondation de l'Automobile Marius Berliet, Lyon, France:** (c, br). **Courtesy of RM Sotheby's:** (tl). **104 From the collections of the American Truck Historical Society:** (tl). **108 Portions of this work are the copyright of PACCAR Inc and is reproduced for limited licensed use only, with permission from PACCAR Inc. Unlicensed use or reproduction are expressly prohibited.:** (tl, cl, bc). **109 From the collections of the American Truck Historical Society:** (ftr, tr). **Gary Alan Nelson:** (ftl, tl). **Portions of this work are the copyright of PACCAR Inc and is reproduced for limited licensed use only, with permission from PACCAR Inc. Unlicensed use or reproduction are expressly prohibited.:** (cr, bc). **110 Dorling Kindersley:** James Mann/John Mould (c). **Gary Alan Nelson:** (cla, tr, bl). **111 Gary Alan Nelson:** (tl, cra, crb, b, cl). **112-113 Mary Evans Picture Library:** Glasshouse Images. **114 Dorling Kindersley:** Milestones Museum (cl). **118 Getty Images:** National Motor Museum/Heritage Images. **119 Alamy Stock Photo:** Humpalova Zuzana/CTK (clb). **Dreamstime.com:** Maria Ivanova (cr). **120-121 Bonhams:** (tc). **120 Copyright © Ford Motor Company. All rights reserved.:** (br). **Gary Alan Nelson:** (cr, clb). **Ben Chr. Sieben:** (cl). **121 Copyright © Ford Motor Company. All rights reserved.:** (b). **From the collections of the American Truck Historical Society:** (tr). **Transport World:** (cl). **122 Gary Alan Nelson:** (tl). **126 Alamy Stock Photo:** CulturalEyes-N (tr). **Bonhams:** (cra). **Gary Alan Nelson:** (bl). **Courtesy of RM Sotheby's:** (cl). **126-127 Bonhams:** (bc). **127 Gary Alan Nelson:** (cra). **Hyman Ltd. www.hymanltd.com:** (tr). **Nissan Heritage Collection:** (tl). **128 Images courtesy of FCA US LLC:** (bl, tr). **129 Alamy Stock Photo:** Mariusz Burcz (br). **Dreamstime.com:** Roman Stasiuk (ftr). **Gary Alan Nelson:** (tl). **Images courtesy of FCA US LLC:** (ftl, tr, c). **130-131 Getty Images:** Sergio del Grande/Mondadori. **132 Alamy Stock Photo:** Oramstock (tr). **Fondation de l'Automobile Marius Berliet, Lyon, France:** (cla). **132-133 Alamy Stock Photo:** Joshua Rainey (b). **133 Alamy Stock Photo:** Paul Christian Gordon (tr). **134 Bonhams:** (cr). **Courtesy of RM Sotheby's:** (cl, bc). **134-35 Courtesy of RM Sotheby's:** (tc). **135 Alamy Stock Photo:** Directphoto Collection (br). **Hyman Ltd. www.hymanltd.com:** (cr, bl). **Courtesy of RM Sotheby's:** (tr, cb). **136 Getty Images:** AFP (tl). **140 Gary Alan Nelson:** (crb). **William Hamilton. Stuart Mitchell:** (cl). **Bob Spear:** (tr). **141 Jay Barnier:** (br). **Bonhams:** (tl). **Nivek Old Gold:** (bl). **Photo courtesy of Mecum Auctions, Inc.:** 2021, Dan Duckworth (c). **142 DAF Trucks N.V.:** (tl, bl, cr). **143 DAF Trucks N.V.:** (ftl, tl, tr, ftr, bc). **144 Alamy Stock Photo:** NPC Collection (tr). **© Granitefan713 (David B):** (cr). **Jack Snell:** (cl). **145 Classic Motors For Sale:** (cr). **Dreamstime.com:** Taina Sohlman (cl); Brian Welker (tl). **NASA:** Dryden photo (tr). **146-147 Dreamstime.com:** Marco Saracco. **148 Getty Images:** Paul Popper/Popperfoto (b); Rolls Press/ Popperfoto (tr). **149 Getty Images:** Dean Conger/Corbis (tr); John van Hasselt/Corbis (br). **150-151 Bonhams:** (br). **150 Alamy Stock Photo:** Goddard Automotive (tr). **Bonhams:** (c). **GAA Classic Cars LLC:** (cl). **Courtesy of RM Sotheby's:** (bl). **151 Alamy Stock Photo:** Matthew Richardson (tl). **Bonhams:** (c). **Getty Images:** Sjoerd van der Wal (cra). **154 Mary Evans Picture Library:** Grenville Collins Postcard Collection. **155 Bridgeman Images:** Peter Edens (ca). **Getty Images:** UPI/ Bettmann Archive (bl). **IVECO SpA:** (cr). **156 Fondation de l'Automobile Marius Berliet, Lyon, France:** (bl, clb). **158 Courtesy of BCVM LEYLAND:** (tl). **162-163 Getty Images:** Rolls Press/Popperfoto. **164 From the collections of the American Truck Historical Society:** (t, bc). **Gary Alan Nelson:** (clb, crb). **165 From the collections of the American Truck Historical Society:** (tc, cra, crb). **Gary Alan Nelson:** (b). **166 Bonhams:** (cl). **Lane Motor Museum, Nashville:** (bl/ Faun Kraka). **Mercedes-Benz Classic:** (tc). **167 Bonhams:** (tr). **Brightwells:** (bc). **Lane Motor Museum, Nashville:** (cl). **Courtesy of RM Sotheby's:** (cr). **Spicers Auctioneers and Valuers:** (tl). **168 Bonhams:** (tr, bl). **Getty Images:** Barrett-Jackson (clb). **Thomas M. Kaercher, Germany, mercedes-pickup.com:** (ca). **169 Bonhams:** (cr). **Dreamstime.com:** Edijs Volcjoks (tr). **From the collections of the American Truck Historical Society:** (cl). **Getty Images/iStock:** Tramino (b). **170 Mercedes-Benz Classic:** (tl, bc, cr). **171 Mercedes-Benz Classic:** (cl, ftl, tl, cb, cr). **172-173 Alamy Stock Photo:** W. Metzen/ClassicStock. **174 Alamy Stock Photo:** Allstar Picture Library Limited (ca); TCD/Prod. DB/Aqua Film Productions (b). **175 Alamy Stock Photo:** Mary Evans/STUDIOCANAL Films Ltd (t); PictureLux/The Hollywood Archive (b). **176 Mercedes-Benz Classic:** (tl). **180-181 Alamy Stock Photo:** Goddard New Era. **182-83 Lane Motor Museum, Nashville:** (c). **182 Alamy Stock Photo: Sentinel3001** (br). **Five Starr Photos:** (bl). **Courtesy of RM Sotheby's:** (tl). **183 Alamy Stock Photo:** Avpics (cr). **Bonhams:** (br). **Nissan Heritage Collection:** (tr). **Photo courtesy of Mecum Auctions, Inc.:** (tl). **184-185 Hyman Ltd. www.hymanltd.com:** (c). **186 Getty Images:** Peter Charlesworth/LightRocket. **187 Getty Images:** Focus/Toomas Tuul/Universal Images Group (clb). **Scania CV AB:** (cr). **188 DAF Trucks N.V.:** (tr). **Volvo Trucks:** (bl). **190 Shutterstock.com:** ITV (tl). **194 Copyright © Ford Motor Company. All rights reserved.:** (tr). **Nissan Heritage Collection:** (cl). **Photo courtesy of Mecum Auctions, Inc.:** (cr). **194-195 Getty Images: Barratt-Jackson** (bc). **195 Fast Lane Classic Cars:** (cl). **Getty Images:** Barratt-Jackson (tr). **Courtesy of RM Sotheby's:** (tl, br). **Jamie Wilson/ jambox998:** (cr). **196 Getty Images:** De Agostini Picture Library (tl). **IVECO SpA:** (c). **196-197 IVECO SpA:** (bc). **197 Alamy Stock Photo:** Frédéric Le Floc'h/DPPI/Gruppo Editoriale LiveMedia (cra). **Dreamstime.com:** Alexander Mirt (cla). **IVECO SpA:** (ftl, tl, tr, ftr). **198-199 Alamy Stock Photo:** Greg Gard. **200 General Motors:** (tl, b). **201 Dreamstime.com:** Artzzz (ftr). **From the collections of the American Truck Historical Society:** (tr). **Gary Alan Nelson:** (ftl, tl). **General Motors:** (br, cla). **202 Alamy Stock Photo:** Simon Bratt (cla). **Getty Images:** Feng Wei Photography (b). **203 Getty Images:** Richard Atrero de Guzman/NurPhoto (t); Asif Hassan/AFP (b). **204 Courtesy of RM Sotheby's:** (tl). **206-207 Alamy Stock Photo:** NG Images. **208 Alamy Stock Photo:** Matthew Richardson (tr). **@car_spots_aus:** (bl). **Dreamstime.com:** Oleg Kovalenko (cra). **Hyman Ltd. www.hymanltd.com:** (cl). **209 Alamy Stock Photo:** Mario Galati (cr); Richard McDowell (b). **Copyright © Ford Motor Company. All rights reserved.:** (tr). **Rob Knight:** (tl). **210 Dreamstime. com:** Yevhenii Volchenkov (cla). **Eric Geisert/EricGeisert.com. 211 Alamy Stock Photo:** Geoff Caddick/PA Images (b). **Getty Images:** Franck Fife/AFP (t). **214 Getty Images:** Jun Sato (c). **215 Alamy Stock Photo:** Erik Pendzich (cr). **Dreamstime.com:** Jack Dudzinski (tc). **216 NY Recovery Ltd. 220 Alamy Stock Photo:** Heather Hacker (bl); Geoff Smith (cl). **Getty Images:** Luis Acosta/ AFP (br). **Wikipedia:** Harry Bush (tc). **221 Alamy Stock Photo:** Avalon/ Construction Photography/BuildPix (bl); Vijit Bagh (br). **Dreamstime.com:** Yevhenii Volchenkov (t). **222 Alamy Stock Photo:** Niccolo Bertoldi (b); Oliver Smart (t). **223 Getty Images:** Souleymane Ag Anara/AFP (b); Jean Guichard/Gamma-Rapho (t). **224-225 Hyman Ltd. www.hymanltd.com:** (c). **224 Alamy Stock Photo:** Dominick Sokotoff/ZUMA Press, Inc. (tl). **Hyman Ltd. www.hymanltd.com:** (cl, bl, bc). **225 Hyman Ltd. www.hymanltd.com:** (tr, ca, cra, c, cr). **226-227 Getty Images:** Sean Gardner. **228 Dreamstime.com:** Jonathan Weiss (bl). **230-231 Shutterstock.com:** Radu Razvan. **232 Scania CV AB:** (tl, cl, cr, bl). **233 Scania CV AB:** (cr, ftl, ftr). **234 Volvo Trucks:** (tl). **238-239 Getty Images:** Zhang Jingang/VCG. **240 Getty Images:** Qilai Shen/Bloomberg (bl). **Shutterstock.com:** Filip Singer/EPA-EFE (tl). **241 Alamy Stock Photo:** Mariusz Burcz (b). **Shutterstock.com. 244 Alamy Stock Photo:** ART Collection (ca). **Dorling Kindersley:** Tuckett Brothers (cra). **Mercedes-Benz Classic:** (crb). **Scania CV AB:** (bl). **Volvo Trucks:** (br). **245 Alamy Stock Photo:** Vitaliy Borushko (tr). **Mercedes-Benz Classic:** (bc). **Shutterstock.com:** Scharfsinn (br). **Volvo Trucks:** (c). **246 Volvo Trucks:** (cr). **247 Alamy Stock Photo:** Scharfsinn (br). **Dreamstime.com:** Timothy Epp (bc). **Volvo Trucks:** (cr).



The publisher would like to thank everyone for their generosity in allowing Dorling Kindersley access to their trucks for photography: Bernard Bailey, Terry Batchelor, Chris Burgess, Simon Carpenter, Alan Dale, Clive Davis, Lee Dunn, Jim Elliott, David Gammond, Ian Gordon, Gregory Distribution Ltd, Gavin Groome, Trevor Howlett, Paul Hutchocks, Stuart Keedwell, Paul Leek, Alan J. Lloyd, M. Way & Son, Dominic Newby, Steve Parker, Mark Purslow, Stephen Reeves, Alan Robinson, James Robinson, RT Keedwell Group, Charles Russell Transport, John Sabin, Pete Sanders, Mark Smith, Ian Snape, Graham Tailby, Richard Thomas, Paul Tunnicliffe, and Neil Yates.

# DEFINITIVE TRANSPORT HISTORY FROM

**THE AIRCRAFT BOOK**
THE DEFINITIVE VISUAL HISTORY
NEW EDITION

ALL-TIME GREATEST AND LATEST AIRCRAFT

**THE BICYCLE BOOK**
THE DEFINITIVE VISUAL HISTORY

ALL-TIME GREATEST AND LATEST BICYCLES

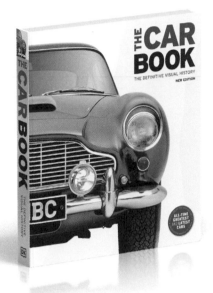

**THE CAR BOOK**
THE DEFINITIVE VISUAL HISTORY
NEW EDITION

ALL-TIME GREATEST AND LATEST CARS

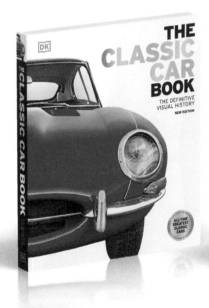

**THE CLASSIC CAR BOOK**
THE DEFINITIVE VISUAL HISTORY
NEW EDITION

ALL-TIME GREATEST CLASSIC CARS

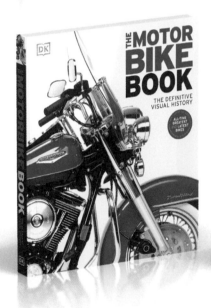

**THE MOTOR BIKE BOOK**
THE DEFINITIVE VISUAL HISTORY

ALL-TIME GREATEST AND LATEST BIKES

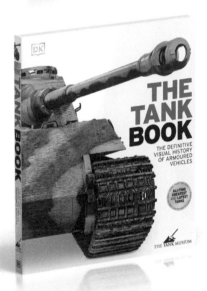

**THE TANK BOOK**
THE DEFINITIVE VISUAL HISTORY OF ARMOURED VEHICLES

ALL-TIME GREATEST AND LATEST TANKS

THE TANK MUSEUM

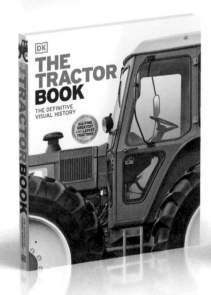

**THE TRACTOR BOOK**
THE DEFINITIVE VISUAL HISTORY

ALL-TIME GREATEST AND LATEST TRACTORS

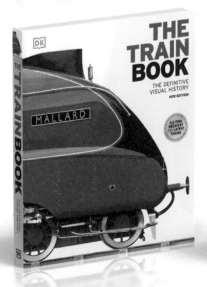

**THE TRAIN BOOK**
THE DEFINITIVE VISUAL HISTORY
NEW EDITION

MALLARD

ALL-TIME GREATEST AND LATEST TRAINS

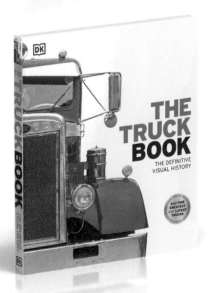

**THE TRUCK BOOK**
THE DEFINITIVE VISUAL HISTORY

ALL-TIME GREATEST AND LATEST TRUCKS